DISCARD

A SHEARWATER BOOK

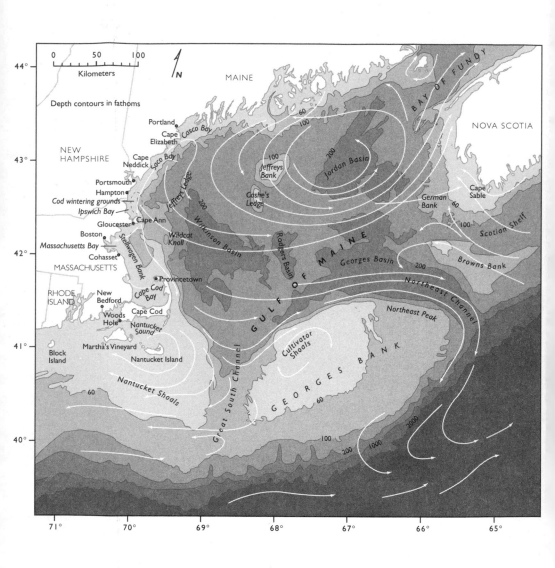

THE
GREAT
GULF

THE
GREAT
GULF

Fishermen, Scientists, and the Struggle
to Revive the World's Greatest Fishery

DAVID DOBBS

ISLAND PRESS / Shearwater Books
Washington, D.C. • Covelo, California

Library of Congress Cataloging-in-Publication Data

Dobbs, David, 1958-
   The great gulf : fishermen, scientists, and the struggle to revive the world's greatest fishery / by David Dobbs.
      p. cm.
Includes bibliographical references (p.   ) and index.
   ISBN 1-55963-663-7 (cloth : acid-free paper)
   1. Fisheries—Maine, Gulf of. 2. Fishery management—Maine, Gulf of.
I. Title.
   SH221.5.M35 D63 2000
   333.75'6'0916345—dc21
                              00-010479

For Alice

If we knew all the laws of Nature, we should need only one fact, or the description of one actual phenomenon, to infer all the particular results at that point. Now we know only a few laws, and our result is vitiated, not, of course, by any confusion or irregularity in Nature, but by our ignorance of essential elements in the calculation. Our notions of law and harmony are commonly confined to those instances which we detect; but the harmony which results from a far greater number of seemingly conflicting, but really concurring, laws, which we have not detected, is still more wonderful. The particular laws are as our points of view, as, to the traveler, a mountain outline varies with every step, and it has an infinite number of profiles, though absolutely but one form. Even when cleft or bored through it is not comprehended in its entireness.

 —Henry David Thoreau, *Walden*

Reason is a supple nymph, and slippery as a fish by nature. She had as leave give her kiss to an absurdity any day, as to syllogistic truth. The absurdity may turn out truer.

 —D. H. Lawrence, "Introduction to the Dragon of the
  Apocalypse by Frederick Carter," in London *Mercury*
  (July 1930); reproduced in *Pheonix: The Posthumous Papers*
  of D. H. Lawrence, part 4, edited by E. McDonald, 1936.

I seem to have been only like a boy playing on the seashore, and diverting myself in now and then finding a smoother pebble or a prettier shell than ordinary, whilst the great ocean of truth lay all undiscovered before me.

 —Sir Isaac Newton, *Memoirs*

Chaos is the law of nature; order is the dream of man.

 —Henry Adams, *The Education of Henry Adams*

O what an endless work have I in hand,
To count the sea's abundant progeny
Whose fruitful seede farre passeth those in land.

 —Edmund Spenser, *The Faerie Queen*

Those who live by the sea can hardly form a single thought of which the sea would not be part.

—Hermann Broch, *The Spell*

The whole of science is nothing more than a refinement of everyday thinking.

—Albert Einstein, *Out of My Later Years*

We love to hear the sayings of old sailors, and their accounts of natural phenomena, which totally ignore, and are ignored by science.

—Henry David Thoreau, *Cape Cod*

Science, which cuts its way through the muddy pond of daily life without mingling with it, casts its wealth to right and left, but the puny boatmen do not know how to fish for it.

—Alexander Herzen, *My Past and Thoughts*

The plural of anecdote is data.

—Nelson Polsby

The sea has never been friendly to man. At most it has been the accomplice of human restlessness.

—Joseph Conrad, *The Mirror of the Sea*

Now mystery masked man was smart,
He got himself a Tonto,
'Cause Tonto did the dirty work for free.
But Tonto he was smarter;
One day said, "Kemosabe:
Kiss my ass, I bought a boat, I'm going out to sea."

—Lyle Lovett, "If I Had a Boat"

CONTENTS

ACKNOWLEDGMENTS

Thanks first, and most vitally, to those who occupy the pages of this book, for opening their lives and work to perusal, a million questions, and the possible trouble that comes of talking to writers, and for sharing with me their thoughts and actions regarding matters close to their hearts and minds: to David and Ellen Goethel, Jay Burnett, Linda Despres, John Galbraith, Steve Murawski, Ted Ames, Dave Crestin, Jorge Barbosa, and Joe and Emily Sinagra. Some of these folks offered hospitality as well—meals, beds, fishing advice, sometimes fishing gear and even fine, sunny-day boat trips *right to the fish,* for which I'm extremely grateful. I can't thank these people enough. Special thanks to the president, cook, and chief bait-keeper of the Wing's Neck Bonito Club, who showed extraordinary hospitality, fish knowledge, and general grace.

I owe many days of safe and stimulating sea travel to Captains Jack Moakley and Jack McAdams of the National Oceanographic and Atmospheric Administration research vessels *Albatross IV* and *Delaware II,* respectively, and their capable, friendly, and accommodating crews. Along with those named above or in the book, I want to thank NOAA fishermen Gene Magan, Tony Alvernaz, Willie Amaro, Doug Roberts, Tony Veiro, and Anthime Brunette; culinary masters Richard Whitehead, Jerome Nelson, and Ernest Foster; bosun and deejay Ken Rondeau; and utility man Orlando Thompson, who provided filleting lessons, peerless company, and showed me the sleeping junco. I took immense pleasure and education, too, from those who sailed with me on the scientific crew as mentors or fellow volunteer scientists: Much thanks for good company and excellent work to Nina Shepherd, Nancy McHugh, Jason Link, Eric Thunberg, Dan Doolittle, Terry Smith, Gavin Begg, and especially my main guides, John Galbraith, Linda Despres, and Jay Burnett. Thanks as well to Tom Azarowitz, head of the Northeast Fisheries Science Center Research Survey Branch, for making room for me on these cruises, and to Northeast Fisheries Science Center director Mike Sissenwine and assessment team scientist Steve Murawski for helping me comprehend NMFS's assessment science.

For aid in understanding the various scientific, environmental, and working perspectives on the Gulf of Maine and Georges Bank fisheries, thanks to fishermen Fred Bennett, John Williamson, Frank Mirarchi, Craig Pendleton, Mark Simonitsch, Mark Farnham, Willie Spear, Mark Leach, Charlie Saunders, Proctor Wells, Terry Stockwell, Tom Brancaleone, James O'Malley, Peter Prybot, Phil Hoysradt, Barbara Stevenson, and Bill Amaru; scientists Rich Langton, Dan Lynch, Greg Lough, Jim Wilson, Dave Mountain, Paul Rago, Michael Fogarty, Robert Beardsley, Eric Thunberg, Gavin Begg, and Trish Clay; and Niaz Dorry of Greenpeace, Jennifer Atkinson of the Conservation Law Foundation, Paul Parker of the Cape Cod Hook Fishermen's Association, and David Bergeron and Angela Sanfilippo of the Gloucester Fishermen's Wives Association. Thanks to Bill Stride for showing me his fish plant and to Vito Calomo for showing me Gloucester and the Gloucester Fish Auction. Some of these people talked to me for hours, others for just a few minutes; all played critical roles in expanding my understanding. Any errors of fact or interpretation are mine.

Several people read and commented on all or significant portions of draft manuscripts of this book, improving it considerably. I'm grateful to Richard Ober, Rick Weston, Madelaine Hall-Arber, Ellie Dorsey, Tim Smith, Ted Ames, and Jim Wilson for this splendid favor.

For help with research and access to materials, thanks to Eva Jonas and Dana Fisher of the Harvard Museum of Comparative Zoology's Ernst Mayr Library; Kathy Norton of the Marine Biological Laboratory/Woods Hole Oceanographic Institute Library; and Janet Nielsen at the Kellogg-Hubbard Library in Montpelier, Vermont. The staffs of the Northeast Fisheries Science Center library and the Woods Hole Oceanographic Institution library annex also helped with access to key materials on fishery science and Henry Bigelow. Anne Bigelow of the Bigelow Society furnished elusive information on Henry Bigelow's family tree.

Kudos and gramercy to Madeleine Hall-Arber of the Massachusetts Institute of Technology Sea Grant program for maintaining the Fishfolk mail list, and to the list's participants for their illuminating discussions on fishery issues.

Thanks to Peter Stein for introducing me to Island Press president Chuck Savitt, and to Chuck for translating an immediate and early enthusiasm for this project into the manna and publishing expertise necessary to make it happen. Thanks, too, to Cecily Kihn for garnering underwriting for the project and to the Derry Foundation and the Maine Community Foundation for their crucial financial support. Editor Dan Sayre cultivated both project and writer with warmth, insight, intelligence, and unflagging enthusiasm. I'm grateful to all the crew at Island Press for their good work.

I owe much to family and friends who provided the support without which no writer thrives. Thanks to Mom, Dad, and Kathy; to Lyn, John, Allen, Ann, Sarah, Jef, Cynthia, Andy, David Allen, Keely, Hallie, and Grace; and to my excellent new family members, Jim, Claudia, and John for giving me the blessings of a boisterous, diverse, and endlessly interesting (!) family. Similar thanks to John and Leita Hancock, Richard and Liz Ober, David Rawson and Karen Freedman, Josh Ober and Adrienne Mayor, Doug Ross and Donna Alterman, David Goodman and Sue Minter, Marylee, Bruce, and Jackie MacDonald and Joe Schwartz and all their clan, Stephen Long, and all my teammates on the Casella softball team.

Finally, I want to thank my dear, fine, lovely son, Taylor Allen Dobbs, who gracefully expressed curiosity about this project, accommodated its inconveniences and distractions, laughed at all my fish jokes, and brings splendid and welcome energy, humor, and intelligence into my world every single day.

And how to thank Alice Anne Colwell? Who indulged a hundred evenings of extra work or escapist diversion; helped me form and reform my thoughts through countless conversations; put on Bach or Billie or Beethoven at all the right times; concocted a head-spinning array of delicious dishes; reviewed and sharpened the entire manuscript; and with similar deftness shapes, refines, and illuminates my life. Thanks, Alice, for recasting my life's ambit.

PROLOGUE

Of the time I spent gathering stories for this book, two experiences return to mind most readily and cleanly, like distillations.

One was watching northern gannets from the bow of the *Albatross IV,* a 187-foot, gleaming white trawler that the National Marine Fisheries Service (NMFS, pronounced "nymphs") uses to track fish populations off coastal New England. Though its name evokes the damning burden of the Coleridge poem, I found the *Albatross* a graceful craft, long of line and blessed with plenty of clean, well-lighted spaces. The cleanest and best lit was the foredeck. While the boat's rear deck, on which the fishing crew landed its hauls and the scientific crew sorted and counted them, was cluttered with gear, the broad, open foredeck was almost featureless, a gently upcurving expanse of pale gray metal. Steel Beach, the scientific crew called it, for its sunbathing qualities in warm weather. In April, however, when I was on the boat, the 10-knot breeze created by our movement across the water kept things chilly, and I usually had the deck to myself. I could lean on the rail at the very tip of the bow, a good three stories above the water, and in solitude look for birds and sea mammals. Sometimes I took my laptop out there and typed up notes, but more often I just stood at the rail with my binoculars and scanned for gannets. I had never been so far at sea—we crossed the entire Gulf of Maine, passing over Georges Bank to within sight of Nova Scotia—and had never seen these soaring pelagic birds. I found them entrancing. Like many birds that spend long periods at sea, gannets have evolved lightweight bodies and broad

1

wingspans that allow them to glide for hours searching for fish. I loved to watch them cruise, floating on their long, canted wings 30 or 40 or even 70 feet up, calmly and alertly eyeing the waves below, then suddenly tucking their wings to drop and spear into the water. They'd pop to the surface a moment later, swallowing, then lift again, their vibrant white bodies framed by black wingtips. They never seemed sullied or ruffled or even particularly wet. Unlike the bickering gulls, they never seemed anxious. It was as if they knew that the sea held what they needed and that they would sooner or later find and effortlessly claim it. I watched them for hours during my weeks on the *Albatross*, slipping off to the bow for at least part of all but the roughest and busiest days. Afterward I always felt a heightened sense of order, even grace.

With similar clarity I recall vomiting off the stern of a much smaller boat. The 44-foot *Ellen Diane* belonged to the fisherman I came to know best, David Goethel. Goethel was in the cabin at the time and kindly took no apparent notice of my Daniel Boone moment (when you go out and shoot your breakfast). I do not generally get seasick, and the seas were not bad that day, maybe 5 or 6 feet, but they rocked the boat just enough that after about an hour on that dank, gray, rolling morning, with too little sleep, too much coffee, and not enough time on small boats, I knew I would soon be sick. I left the cabin, went back onto the fish deck, clutched the strut of the net transom so I wouldn't be tossed overboard, and leaned over the rail. Each roll of the boat brought the water almost within arm's reach. I stared at the sliding sea, intentionally letting its swinging, surging passage accelerate the inevitable. Finally I surrendered to my stomach's insistence. Emphatically. I hadn't vomited that violently since, oh, college. I had forgotten how the muscles that lace the abdomen can clench so hard around your stomach. I was rattled by how much it hurt and relieved when it stopped. Along with watching the gannets, this experience too left me feeling better and more ordered, though well short of grace. It was a different kind of order—a return to a tired but whole state where I figured I could get through the day after all. I felt no sense of elevation, just as there was no sense of elevation above the water in Goethel's boat. I stood awhile, holding the transom and watching the sky and horizon in hopes of seeing birds or whales or dolphins, something to divert my attention. But they had fled elsewhere on this scouring day.

I went back inside, my face damp with sea mist, and talked some more with Dave Goethel about how he and so many other fishermen could see this stretch of water so differently than the National Marine Fisheries Service saw it.

I had become fascinated with a squabble that took its most visible form in a disagreement, by orders of magnitude, over how many fish were in this piece of ocean. The argument seemed astonishing or even ludicrous when you first encountered it. How could two groups of people who spent so much time on the same water differ so profoundly on the population count of its major species? Granted, the sea is opaque. Counting fish is more challenging than taking census of caribou or bears or coyotes or even songbirds. Still, I initially found it incredible, in every sense of that word, that two well-informed parties could not agree on the status of some of the system's biggest, most dominant species. Yet that was the case. The fishermen and NMFS scientists of the New England fishery, our country's oldest, most important, and most thoroughly studied, argued about the health of the cod and haddock stocks of Georges Bank for over a decade until those stocks collapsed in 1992, and ever since then they had been arguing about the cod population of the inner Gulf of Maine, which at the century's close was, depending on whom you asked, either in immediate danger of complete collapse or doing fairly well.

Cod composed the heart of the New England groundfishery, which had for centuries been one of the world's most productive fisheries. It was the fish, as described in Mark Kurlansky's *Cod: A Biography of the Fish That Changed the World,* that altered history by so enriching the colonists they quickly became major players in international trade.[1] When European settlers first came to New England, they found cod in an "abundance," wrote one Salem minister in 1629, "almost beyond believing." The fish surged along the coast by the tens of thousands, some as big as grown men. They readily took the hook. Worldwide demand for their flesh, which tastes as clean as its snowy white appearance suggests, soon made the colonies an increasingly self-sufficient, self-confident political power and gave them the financial means to seek and win political independence.

Over the decades that followed, Americans would find more astonishingly abundant resources—timber, coal, farmland, oil. It was with the New England cod, however, that this young nation devised what a biologist might call its life strategy: Find nature's richness, then exploit and defend it with ferocious independence. In a country where seemingly endless natural blessings created unprecedented wealth and a singularly individualistic optimism, the cod of New England had been the first such blessing.

Now we had decimated this richness. We had overfished New England's waters so severely that some scientists worried we had destroyed something vital. And amid the ensuing disbelief and anger raged this curious battle between fishermen and scientists.[*] In its most simplistic and, unfortunately, most common expression, their dispute took the form of a shouting match—"It's your fault!" "No, it's yours!"—which made almost everyone appear foolish, callous, or both. The whole thing made you suspect that those on one of the sides were deluding themselves. The gigabytes of scientific evidence marshaled by NMFS, of course, suggested that the deluded group was the fishermen.

What was going on here? Part of the answer lies within the two memories I described above: one of an apparently effortless, relatively detached observation from a high platform with views far and wide (and an occasional dip to the surface); the other of an engagement with the ocean that, if it lacks the high perspective of the first, is far more direct, visceral, and intimate. As quite a few fishermen and scientists have noted (they can agree at least on this), NMFS, dropping its sampling nets all over the Gulf of Maine and Georges Bank twice yearly, has a better view of the big picture than most fishermen do, while the fishermen, out on the ocean every

---

[*]The New England fishery, though one of the worst, is far from unique in its failure to curb overfishing in the past few decades. As the twenty-first century opened, eleven of the world's fifteen most important marine fishing areas were in decline, and 60 percent of commercial fish species were being fished at or beyond capacity. In the United States alone, mismanagement and overfishing, often aggravated by disputes over science similar to the controversy in New England, had seriously depleted major fisheries on all three shores—the Gulf of Mexico, the Pacific, and most of the Atlantic coast. Similar problems affected at least one if not most major fisheries on every continent, causing incalculable ecological, social, and economic disruption.

day, know any given piece of water in far more detail and under many more conditions and circumstances.

In an ideal world, these two views would merge into something richer. But they had not, and the more I looked at the New England fishery, the more I believed the failure to reconcile these two perspectives had sped the fishery's collapse and was crippling the effort to revive it.

I do not mean to substitute the obscure for the obvious: Overfishing caused the fishery's collapse, and overfishing threatens its recovery. Nor do I mean to imply that the National Marine Fisheries Service failed to generate the scientific knowledge necessary to warn of and correct that overfishing. The service had begun warning of overfishing in the mid-1980s, and had the New England Fishery Management Council (the federally authorized regulatory body, composed of state fishery regulators and fishermen appointed by the region's governors, that was supposed to regulate the fishery) responded to that information by imposing reasonable restraints, it could have prevented the whole mess with much less sacrifice by those who fished. Today we'd still have plenty of fish and happy fishermen alike.

As I hope this book makes clear, the rift between fishermen and NMFS scientists over how to look at the ocean and think about fish fostered a level of discord, doubt, and mistrust that made it almost impossible to convince fishermen and regulators to curb overfishing. It also impoverished our understanding of the ocean, including NMFS's assessments of fish populations, by hindering the exchange of information and perspective that might have created a fuller, more informed, and more informative science. Science that respected and included the vast knowledge held by the fishing community would have stood a far better chance of persuading fishermen to restrain their fishing. And today it would provide the additional insight we need to revive the fishery.

How did this schism come to exist? In the late nineteenth century, when NMFS's predecessor, the U.S. Fish Commission, was established, the two parties had started out committed to a single, mutual cause, a healthy fishery. But after years of bickering and fading commitment, they found themselves, much like an estranged couple, first in a state of alienation and then enmity, suspicion, even hatred. It had never been an easy union. It had begun with promise, however, the two sides dedicated if not to each other then to a shared devotion to the sea and its health: the government scientists to the monitoring and care of the fishery, the fishermen to a respect for a science that aided them. Now the two sides could scarcely speak to each other. What went wrong? Who strayed? Who first abandoned their vows?

Seen that way, the dispute seems explainable only by attributing blindness, dishonesty, or insanity to one side or the other—which many did. "Who Says There's No Fish?" was a popular bumper sticker in fishing ports in the 1990s, and the defiant antagonism in that statement speaks volumes about how deeply some fishermen were willing to look at the problem and about how much they trusted NMFS's numbers. At the other end of the spectrum, some scientists believed that most of the fishermen were either lying or willfully myopic when they insisted there were plenty of fish.

This haggling over fish numbers, however, was merely the distractingly visible part of the dispute—a sort of red herring, if you will, made necessary and critical only because the New England Fishery Management Council's consideration of NMFS's population assessments offered the only place where the opinions of fishermen and scientists regarding fish populations ever met. By that time, of course, it was far too late to reconcile the two perspectives; that could be accomplished only much further up the perceptual stream, so to speak, and nothing in the regulatory set-up encouraged such convergence. NMFS scientists and the fishermen thus ended up arguing over numbers, and each side said things that seemed inexplicable to the other.

Yet if you go to sea with some of the many sensible, conscientious, and attentive people on either side of this schism, you find they are not fools or boors and that their disagreement is not simply a spat between parties blinded by rage or arrogance (though it's in constant danger of so becoming). Rather it is a disagreement among knowledgeable, honest people about fundamental definitions of fact, perception, and understanding. That two parties who know the same water so well can regard it so differently is not an absurdity; it's a Sphinxian riddle, with the solution illuminating not only the complexities of a fascinating science and an extraordinarily fertile piece of the earth but also the difficulties we all face in trying to see the world as freshly and completely as possible. For what is a Sphinxian riddle if not a challenge to our usual habits of thought? This puzzle of how many fish were in the ocean is partly an argument about fish, science, and culture; it's also an argument about how to view the world.

The dispute between fishermen and scientists draws intensity from the cultural chasm between them, which mirrors a gap that regularly hinders the efforts of society at large to resolve conflicts about natural resource extraction—that is, those battles regarding resources most of us use, such as timber, oil, and fish, that are pulled from the earth by some subset of our society to whom we hand this work. Conflicts like these tend to get cast in the well-worn "environmental issue" template, a loggers-versus-Sierra-Club script, one party taking another to task for "abusing the resource" or "raping the earth," the other clumsily, often righteously, defending itself. I find this the dullest and least productive way to approach such problems. It stages as a tired morality play, good guys against bad, a conflict that is essentially about our alienation from the earth (and here I mean the dirty part of the earth and its creatures, the crumbly and itchy and slimy stuff, the soil and mud and blood and mucus and milt and roe that we wash from ourselves after we engage life especially closely) and especially from the task of yanking from the earth our sustenance and (so embarrassing to consider!) our wealth. I had already seen, while writing with a friend a book about people who earn their livings from New England's forest, how readily the softer-living among us take offense at this extractive work and how we express our anxiety and horror by blaming the people who have taken or been stuck with the roles of direct extractors.[2] We cast them as villains, projecting onto them not just the physical but also the moral and ethical responsibility for ripping up the raw materials for our endless needs and wants.

This communal alienation and discomfort seems distinctly at play in this fish fight, both in the rift between fishermen and scientists and in the larger public's readiness to accept the easy answer that the fishermen just got too greedy. At play as well, and helping to explain the apparently contradictory nostalgia and guarded respect people feel for "good" fishermen, is the subdued, largely unconscious regret that most of us feel about our own lack of deep physical engagement with the natural world. This play of interwoven currents and tensions, rather than the more audible uproar over fish numbers or the battle between the catch-'em-all and save-'em-all camps, is what caught me up in this story, and is what I attempt to follow in these pages.

Resolving these tensions and divisions is difficult, and it strikes me that it almost has to be done more on the water than in meeting rooms. This was the approach taken by the scientist who merged the perspectives of fishermen and scientists most successfully, an avid angler, sailor, zoologist, and oceanographer named Henry Bryant Bigelow. The most accomplished oceanographer and fishery scientist of his generation, Bigelow pioneered the scientific exploration of the Gulf of Maine in the early twentieth century. He came to know the Gulf perhaps as well as anyone has ever known any large natural system, and he did so largely by blending the mental habits and knowledge of the scientist and the fisherman. This integration made his work, particularly his comprehensive field guide *Fishes of the Gulf of Maine,* so informative and inspiring that even today he is admired unreservedly by both sides. No individual today could command the sort of comprehensive knowledge that Bigelow possessed. There is simply too much to know. Yet there is no reason Bigelow's determinedly inclusive, synthesizing approach can't be more common. This book is, among other things, a call to let the Henry Bigelow in scientists and fishermen come out and run things.

Beyond that premise, I do not pretend to offer some precise lesson about how to rescue the fishery or live better or save our souls through closer connections to the earth. We cannot all go into fishing or farming or logging (though I believe we can live and behave more decently if we understand those who do).

I don't want to make this rather exciting dispute over fish and fishery science seem too somber. The story of this schism and its effects clearly has sobering aspects. Yet it is marvelously and inspiringly full of life—on one hand the vitality of an uncontrollable, ultimately unknowable sea and its strange, frightening, and beautiful creatures, and on the other the energy and spirit of smart, irrepressible, unpredictable people who work there with great joy and humor and who refuse to surrender to despair or cynicism. I found here some of the funniest, most generous, and most fascinating people I've ever met—the more so, strangely, for being caught in the "God-awful time," as one of them put it, of today's fisheries crisis.

It's a mess these angry, confused, and determined people are still trying to sort out (aided, or not, rather clumsily by the rest of us), and it's far from clear how it will end. Despite the considerable damage to the Gulf of Maine and Georges Bank, I don't worry in any essential, absolute sense about how those biological systems will fare in the long term. My guess is that by means either ugly (see part 4) or more graceful (see part 5 and the epilogue), we will restrain our fishing enough to let these fecund waters recover. I do worry, however, whether in trying to save this and other fisheries we will leave intact any working relationship between the people of our coastal towns and our ocean, something more vital and profoundly cultural than aesthetic or recreational connections. I wonder, too, whether fishermen and scientists (and the rest of us, for that matter) can reconcile the various ways we see the natural world to produce a more nuanced, inclusive, and dynamic perspective—a view that takes fuller account of nature human and otherwise.

In my most optimistic moments, I believe these changes can occur. You can make a strong argument that all of it—the fish, the fishermen, the ocean, the effort to see better and to enrich our communal and individual relations to the outdoors—will tank. When you consider everything pushing us that way, it's easy to succumb to despair or apathy. But if you come to know a few people who see things otherwise, you tend to take heart at the good acts they are capable of. Next thing you know, you feel hope.

Whether such hope is well founded, who knows? Yet each day we rise.

PART I

I

On July 9, 1912, zoologist and oceanographer Henry Bryant Bigelow sailed a 90-foot schooner, the *Grampus,* out of Gloucester Harbor and onto the Gulf of Maine. The *Grampus* belonged to the U.S. Fisheries Bureau, which had hired Bigelow at a dollar a day to find out why fish catches in the Gulf of Maine were dropping.

Fine with Bigelow. An expert sailor and insatiable outdoorsman, the thirty-two-year-old had grown restless in his day job—cataloguing jellyfish, anemones, and tiny hydrozoans at the Harvard Museum of Comparative Zoology—and was glad to get out on the water. As a young scientist looking to make his mark, he also was happy to use a government boat to conduct an original investigation of an important but relatively unstudied piece of water. Later, writing one of his reports, he recalled feeling that "few living zoologists have been as fortunately placed as were we on setting sail on the *Grampus . . .* , for a veritable *mare incognitum* lay before us."

Mare incognitum: a sea unexamined. Bigelow set out to examine this one with thermometers and sampling bottles, a half-dozen plankton nets, a trawl net, a dredge, some drift bottles, and a 30-pound weight and some plumb line. His venture sounds quite pioneering and in many ways was. Bigelow's long investigation of the Gulf of Maine stands as one of the most monumental accomplishments in modern oceanography and one of the more impressive pieces of natural science fieldwork. Spending much of the 1910s and early 1920s cruising the Gulf, he almost single-handedly turned its great wet oblong from a scientific unknown into the world's best-described large marine ecosystem. He answered scores of vital, central questions about the Gulf and its sea life, produced in *Fishes of the Gulf of Maine* one of the most insightful and enduring books about fish ever written, and propelled himself to the pinnacle of his field.[1]

Bigelow's notion of the Gulf of Maine as a *mare incognitum* was a conceit, however, and Bigelow knew it. Fishermen had been working the Gulf for four centuries when Bigelow declared it unexamined, and the best of them knew the water intimately and in detail. Though they might not take temperatures, measure currents, sound depths, and catalogue species, they attended closely to the movement of fish and to the kaleidoscopic swirl of variables—tide, temperature, wind, weather, current—that seemed to push fish in and out of feeding grounds and to make them plentiful in some years and scarce in others. Unlike scientists, the fishermen did not quantify and categorize this information into any sort of shareable database. They added it a drop at a time to pools of knowledge, the particles cohering to each other like water molecules, on which they drew instinctively. They didn't know data; they knew the sea.

By the time Bigelow sailed the *Grampus* out of Gloucester, in fact, the fishermen of the Gulf of Maine knew the sea well enough to have fished the halibut, one of the Gulf's most prolific species, into near extinction and to have put increasing strain on cod and haddock populations that had long seemed limitless. The overfishing led the Bureau of Fisheries[*] to hire Bigelow and lend him its boat. His job, in essence, was to quantify some of what the fishermen knew, such as what fish were out there and where and how they lived, and to discover some of the things the fishermen didn't know, such as subsurface temperatures, salinity, precise depths, intricacies of current, and the causes of plankton abundance.

Though he would become more and more interested in integrating these two areas of knowledge, Bigelow began the project focusing mainly on "oceanographic" facts, and he agreed to collect information on fish as part of the price of getting the Fisheries Bureau's sponsorship. In particular he wanted to discern what dynamics of current and hydrography made the Gulf's waters so cool (cooler by some 5 to 10 degrees centigrade, or 9 to 18 degrees Fahrenheit) than the water just outside its gates. He was convinced that those oceanographic dynamics would, in turn, explain why the

---

[*]The Bureau of Fisheries began as the independent U.S. Fish Commission in 1871. It changed its name to the Bureau of Fisheries when it was put under the Department of Commerce in 1903, and became the National Marine Fisheries Service in 1970.

Gulf and Georges Bank, the underwater plateau that forms the Gulf's southern rim, support such tremendous numbers of fish.

Bigelow's accomplishment in the Gulf was by any measure astounding. He answered many of the biggest and most important questions about the Gulf's currents, oceanography, and fecundity. He catalogued and vividly described every fish found there. And in parsing the whirled workings of the Gulf and its bordering waters, Bigelow did more than get a handle on some of the United States' richest fishing grounds. In his insistence on drawing connections, on making the jump from "the mere accumulation of facts," as he put it, to an approach he called "the fitting of these facts together," he moved American oceanography from the specimen-collecting days of the nineteenth century to the ecosystem-oriented perspective of the twentieth.[2]

He also bound together, for a brief, graceful moment, the seemingly unmeshable angles of view of the scientist and the fisherman, merging the two, like the twin images from a set of binoculars, into a deeper, more detailed and perspective-rich picture than either could provide alone. He did not set out to do this. His writings and letters hold little to suggest he saw the schism between scientists and fishermen as some destructive social or political gap that needed to be closed. He did it because he needed the information that fishermen possessed. But the result was something rare and valuable: a union of the intuitively piscatory and the self-consciously scientific into an integrated, dynamic, and superbly productive understanding.

Although Henry Bigelow is widely acknowledged as the most influential figure in American oceanography, the synthesis of approach he achieved in his work on the Gulf of Maine and Georges Bank, which he captured brilliantly in the three books he wrote in the 1920s, was largely lost in the decades that followed.

When he sailed from Gloucester that July day in 1912, Bigelow was seeking merely to get a scientific handle on the Gulf of Maine—and to get out of the lab. He loved being outdoors as few have. Though not a diarist or particularly introspective as a letter writer, he did write a short memoir in

1964, when he was eighty-five. This strange, rambling, artlessly charming forty-one-page document portrays someone who from boyhood did his scholarly work well but rose to his highest levels of energy and innovation outside.

He grew up in Boston, the son of a prosperous banker, Joseph Bigelow, the second of six children in a family that stressed a broad, rigorous education and an active life. He attended Milton Academy and then Harvard, both undergraduate and graduate school, and traveled to Europe several times in his teens. In its broadest outlines—the prep school background, the Harvard lineage, the European tours, the summer house on the shore—his life seemed that of a slightly decadent Boston dilettante. His energy and curiosity, though, allowed no decay, and he never cared to dabble. He *did.*

His happiest and most formative youthful experiences were at the family's summer house in Cohasset, on the coast of Massachusetts Bay, 15 miles southeast of Boston. There he learned skills he would hone throughout his life—sailing, canoeing, navigating, learning to read water and weather, and zealously fishing and hunting.

He also took an early lesson in using the knowledge of fishermen, though in pursuit of his personal fishing agenda rather than science. His father (an indulgent man but one who demanded industry and discipline from his children) one summer suggested that Henry and his older brother, Joseph, earn some money, perhaps by tending the family's large vegetable garden. Instead, Henry and Joseph did as many coastal New England adolescents have done and still do: They went lobstering. On breezy days, the boys sailed the family's 17-foot sailboat, the *Shrimp*, from pot to pot; on calm days, they poled with a long sculling oar. The project "was a financial success from the start," Bigelow recalled seventy years later, "partly because we were familiar with the local waters, but perhaps chiefly because the local lobster fishermen, who held our father in high regard, helped us in many ways." A classic Bigelow synthesis: From the basis of his privileged upbringing, he wove his own observations and the fishermen's knowledge into a successful undertaking.

Bigelow's memoirs are packed so full of recollections of adventure that, as a colleague once noted, you wonder when the man worked. He climbed in Switzerland and British Columbia; skied in Switzerland, Austria, New England, and British Columbia; canoed, snowshoed, and hiked all over New England, Newfoundland, and New Brunswick; fished throughout the United States, Canada, and Europe; and hunted all sorts of game in North America. At the end of his first big oceanographic trip, in his early twenties, he wrote a letter home from Ceylon (now Sri Lanka), describing how he had bagged "two wild water buffaloes, a sambhar stag, two axis deer, one wild boar, an eight-foot crocodile, and a large python."[3]

His passion for adventure rose early. The money he and his brother earned lobstering, for instance, they spent on shotguns so they could learn from an uncle the

> art of rigging out for "sea coot" shooting, for which both of us developed a great fondness. True, it was a hard sport, for it involved getting up before daylight, rowing out into the open sea for a mile and a half or so, and anchoring sometimes in very rough weather so that the shooting was the most difficult I have ever done. The rewards, however, were great visually as well as in the bag, for on a good day, skein after skein of fowl was passing by, so that hundreds of fowl were often visible at one time.[4]

Bigelow always loved to hunt, and in his twenties and thirties he was spurred on by the scientific value of the more exotic game as specimens for the Museum of Comparative Zoology, where he studied and worked his entire adult life. (He once acquired a porcupine for the museum by shooting a branch out from under it.) He loved hunting for the sport itself and because it meant being outdoors in the most active way possible. In his later years, for instance, he preferred hunting upland birds to waterfowl because to hunt upland birds, you must walk.

Always determinedly practical, he took pleasure in knowing how to cook the game he shot. He tells us that breast of sea coot is best broiled over a hot fire and served with a sauce of wine and currant jelly, and that muskrat is tasty if you remove the musk glands before cooking it. At least once he fashioned game skins into clothing. On his honeymoon, a two-week canoe trip into the Newfoundland wilderness in 1906, his wife, Elizabeth Perkins Shattuck, lost most of her clothes, including her shoes, when the canoe she was sharing with one of the guides capsized. Bigelow built her a fire, shouldered his rifle, hiked a mile upstream, found and shot a caribou, then transformed the skin of its legs into a pair of boots for his bride.*

This was not a man who was going to do his best work indoors. It's probably impossible to overestimate what Bigelow's physical restlessness, his compulsion to engage the world with his body as well as his mind, brought to his work—or, for that matter, how lucky he was to find work that let him exercise that compulsion. He loved oceanography's physical and logistical challenges—the sailing and navigation, the duct-tape engineering, the invention or modification of gear, the weather, the work on deck. These enthusiasms made him a better oceanographer, for they encouraged him to go out repeatedly to collect data. His perpetual engagement in such practical concerns also added rigor to his theoretical side. Like the discipline itself, the fieldwork of oceanography required imposing on the natural world an intellectual order that was truly cognizant of the sea's dynamic nature. A navigational decision that sets you on a certain course, the repair of a piece of dredging equipment that must sink fathoms deep and scoop up sea bottom, the design of a drift bottle that will move easily with the current yet survive all weather and seas—these things were, in essence, theories that would be tested mercilessly. If you send a weak idea into the sea, it will almost certainly come back mangled. To constantly check one's ideas against such demands, and to experience the ocean so often and under so many conditions that it provides the texture and patterns of one's consciousness, can't help but encourage a reality-based style of theoretical inquiry.

Bigelow first tasted these pleasures the summer before his senior year in college, in 1900, when he joined a Harvard and a Brown professor and four other Harvard undergrads for two months of geological, botanical, and zoological investigations along the Newfoundland and Labrador

---

*His memoir includes instructions for doing this—presumably in case the reader should have similar need: "One cuts the skin of the leg across far enough above the hock for the upper part to serve as the leg of the boot and far enough below the hock to accommodate one's foot and leg. It is desirable, furthermore, to tan the boots to prevent their inner surface from continuing disgustingly slimy." This he accomplished by boiling cherry and maple bark in a solution of baking powder and soaking the boots in it overnight.

coasts. The trip offered an ideal combination of adventure, tomfoolery, and scientific work. Bigelow took great delight in the Eskimo sealskin boots they were given, and he seemed pleased that the ship carried no fresh meat, making it necessary to do a little shooting ("some young gulls, which are delicious, and sea pigeons [black guillemots], which are at least edible").[5] They also caught cod by hook and capelin by cast-net. The boat maneuvered amid ice floes and bergs, unusually heavy that year, inspiring in Bigelow a long fascination with the problems of navigation in ice.[*] He paid particular attention to the birds that nest so densely in the Canadian Maritime: black guillemots and razor-billed auks skimming the waters, puffins plunging off the cliffs, and scores of other shorebird and waterfowl species nesting and feeding. Bigelow took enough notes on birds to write one of his first published scientific papers, a simple but fairly comprehensive "Birds of the Northeastern Coast of Labrador," which appeared in 1902 in the respected ornithological journal *The Auk*.[6]

Even after this trip, however, Bigelow was uncertain about whether to go into oceanography. That hook was set deeply a year later, when he took the first of many voyages with Alexander Agassiz, the director of Harvard's Museum of Comparative Zoology.

2

Bigelow did not know Alexander Agassiz before he signed up for his 1901–1902 Pacific voyage, as Agassiz, spared any teaching duties, rarely crossed paths with undergraduates. But Bigelow had seen him around the museum, where Agassiz and his huge reputation were unavoidable. He was the preeminent oceanographer of his time and country, and perhaps the

---

[*]In the 1920s, in the aftermath of the *Titanic* sinking, Bigelow served as a scientific advisor to the International Ice Patrol, which created guidelines for tracking and avoiding the icebergs that calve off of Greenland and drift south.

world; his only competition for that latter distinction, his Scottish friend Sir John Murray, wrote after Agassiz's death that "the present state of our knowledge . . . is due . . . more to the work and inspiration of Alexander Agassiz than to any other single man."[7]

The Agassiz name loomed all the larger at Harvard, for Alexander's father, Louis, had founded Harvard's Museum of Comparative Zoology in the 1850s, and he and Alexander had made the museum one of the country's leading scientific institutions. A great popularizer of natural history, the elder Agassiz had come to the United States from his native Switzerland for a short lecture visit in 1846 and so enthralled Cambridge's intelligentsia with his lectures that Harvard offered him a professorship. He started the museum a few years later.

Alexander shared his father's brains and charm but lacked his extroverted personality. He hated teaching and avoided attention. Yet he was a more thorough scientist, a more organized administrator, and a far better money manager than his father—the perfect choice to strengthen the museum's foothold in the scientific world. Alexander, who had advanced degrees in both natural history and engineering, made a fortune by solving some of the technical problems of a copper mine in which he owned an interest, and over the course of his lifetime he donated more than $1.5 million to the museum. He also drew deftly on his and his father's reputations to raise funds. He was thus able to give the museum, which in his father's day had stumbled from one funding crisis to another, a firm financial foundation.

His riches also allowed him to lead some of the most extensive oceanographic expeditions in history. He sailed for three to six months almost every winter of his life beginning in the mid-1870s, first in search of sea urchins, on which he became the world's leading expert in his thirties, then to examine coral reefs. His coral reef investigations took him all over the Caribbean and the Pacific, an itinerary that kept him busy almost until his death, in 1910.

It was on one of Agassiz's last Pacific expeditions that Bigelow first joined him, the fall after he graduated from Harvard. Toward the end of his senior year, Bigelow heard that the famous director was planning yet another trip to exotic isles, and, inspired by his shorter journey to Newfoundland the previous summer, he introduced himself and offered his services as a research assistant. Agassiz, who must have been delighted to enlist an assistant so bright, energetic, and sea-savvy, signed him up.

Their destination was the Maldives, a large group of coral islands 200 miles off India's southern tip. They sailed first, via the Atlantic, the Mediterranean, the Suez Canal, and the Red and Arabian Seas, to Ceylon, a large island just off the Indian peninsula, where they picked up the *Amra*, a comfortable British passenger steamer Agassiz had chartered. Agassiz, a stickler for advance preparation, was pleased to find that the modifications he had ordered for the ship had been carried out nicely. The ship's engineers had installed a depth-sounding machine, a dredging winch, and all the supplies necessary for a first-class on-board laboratory for processing, preserving, and analyzing biological and coral specimens.

In the Maldives, the only major coral formation he had not yet visited, Agassiz hoped to complete the picture of coral reef geology he had been accumulating over the previous twenty-five years. He also wanted, as always, to fully sample the sea life, particularly the invertebrate life, and it was for this purpose that he took along assistants such as Bigelow. He assigned Bigelow the job of collecting and preserving for future study all the medusae, or jellyfish, that the ship's plankton hauls would draw up. Bigelow was also to help sort the many other life forms the dips into the ocean fetched up. Agassiz, Bigelow, and the rest of the scientific crew sampled the sea life by dredging (dragging a heavy scoop along the bottom), to obtain samples of the sea bottom and the life that clung to it; by towing through the middle and upper layers of water a wide-mouthed net of finely meshed silk that caught the tiny floating plants and animals that composed the area's plankton; and occasionally by lowering a fishing net or baited lines to catch fish. They also regularly lowered instruments into the water to measure its temperature, depth, and salinity.

Sampling and monitoring procedures were then and still are the staples of any oceanographic outing; Bigelow was lucky to learn the techniques from one of the most meticulous, disciplined, and logistically skilled field scientists who ever lived. Bigelow witnessed the value of applying engineering knowledge to the design of oceanographic equipment, for Agassiz was masterful at modifying, repairing, and even inventing dredges, thermometers, trawl nets, and other sampling equipment. Bigelow also learned from Agassiz how to set up and monitor the processing and storage of specimens; how to train a ship's crew to take the care required for accurate work; and how to plan an efficient day's work and trade off navigational against scientific considerations. Perhaps most valuable, he learned the importance of doing solid, thorough work even when bad weather, rough seas, fatigue, homesickness, seasickness, equipment trouble, or other discouragements made you want to compromise.

Not that all was drudgery. The trip offered the diversions and novelty of tropical waters and their colorful fish, as well as fine twilit dinners, complete with wine, beneath the awning on the ship's broad wooden deck. It also offered company with Agassiz, whose charm and intellectual example only grew on those who traveled with him. Witnessing Agassiz's unquenchable and infectious curiosity was one of the great pleasures and lessons Bigelow took from this and subsequent cruises. Agassiz was said to be intrigued by every one of the thousands of hauls he ever saw swung aboard. He was present almost every time a sampling apparatus was emptied, eager to sort through the catch and sluice into his favorite viewing instrument, a large glass bowl set over a bright light, a few cupfuls of the sea water so he could see what small invertebrate life it held. Once, too ill with flu to climb back and forth between his stateroom and the deck, he slept in a lounge chair up top so he could monitor the dredge as it worked through the night. This inexhaustible curiosity did much to make him the great teacher, cruise leader, and shipmate he was. It taught that every sample held scientific importance and stimulated an interest in operations that otherwise could become deadening. The repetitive tasks, often carried out late into the night and under taxing conditions, can turn oppressive if you cease to care what comes out of the water. Once that interest is lost, every haul is simply work, every sampling station another obstacle to returning home, and the homesickness that stalks everyone on any long ocean cruise takes over.

Agassiz's disciplined inquisitiveness both devolved from and fed his ferocious insistence that accurate field observation forms the bedrock of good science. Abundant data derived from careful fieldwork, of course, do indeed underlie most scientific advances. Agassiz, though, seemed to take the notion to almost obsessive ends. "Study life, not books," his father had famously stressed, and while neither Agassiz underestimated the importance of reading deeply and widely, Alexander heeded perhaps too fully this maxim's implicit elevation of observation over theory. He never liked to theorize about a subject unless he commanded a wide body of observed data about it. He thought those willing to do otherwise in private were sloppy and those willing to do so in public were dangerous.

This obsession prevented him from ever writing up or apparently even forming a coherent theory about coral reef formation, though he studied reefs for more than thirty years. Any organized thoughts he had on the subject were lost in March 1910, when he died quietly in his bunk while crossing the Atlantic from Europe, where he had gone to confer with colleagues and examine once more some coral reef collections. His papers left no clue as to his thinking. As his son's account gracefully puts it, "The material he left furnishes an excellent example of his method of carrying his work in his head until the last moment. At his death nothing could be found but a vast collection of extracts from the literature of the subject marked and scored, and a few rough notes, of no use to any one but himself."[8]

Henry Bigelow witnessed most of the last chapters of this saga, traveling with Agassiz on another Pacific journey in 1904 and to the West Indies in 1907 for yet another look at the reefs there. What Bigelow thought of Agassiz's failure to write up his theory he never said. Yet it seems clear that while Bigelow absorbed Agassiz's dedication to thorough observation, he did not share Agassiz's reluctance to theorize about those observations. He recognized the perils of letting the perfect be the enemy of the good—or, to put a more practical spin on it, he saw the wisdom of E. B. White's maxim, "Get it right, but get it written." Whereas Agassiz seemed happy to collect facts endlessly without connecting them in any extensive way, Bigelow gathered facts—making quite sure he had as many as were practical to gather—so he could find the vital connections among them. As he put it in a 1930 essay, "The mere accumulation of facts from the sea, when there is an inexhaustible supply, may become a bit sterile, just as catching fish is to a sportsman where fish are too plentiful. . . . What is really interesting in sea science is the fitting of these facts together."[9] He had the same judgment and powers of synthesis regarding his mentor as he did about the dynamics of the Gulf of Maine.

3

Bigelow had other teachers, of course, and plenty to keep him busy between trips with Agassiz. Along with writing reports on the medusae and siphonophores he had collected on his Agassiz voyages, he worked on a Ph.D., which he completed in 1906 with a dissertation on the nuclear cycle of the medusa *Gonionemus murbachii.* This academic work taught him the value of careful laboratory procedure and of reading widely and thoroughly. He learned this latter lesson the hard way: After he had selected a subject for his master's thesis, his thesis advisor, Dr. Mark, reminded him to look up all the literature on the subject. As he noted in his memoirs,

> I replied that I had done so. This . . . was in August. In November Dr. Mark rather casually suggested that perhaps I'd better look up a certain paper that had been published in Austria in the 1870s. In doing so I found that someone had forestalled me. When I took Dr. Mark to task for having kept me so long in ignorance of this he replied, "Henry, the next time you look up the literature on any subject, you'll look it up," and I have done so.[10]

Bigelow taught such lessons as well. Once, when an undergrad included in an anatomical drawing of an invertebrate several organs that only vertebrates possess, Bigelow so severely berated him for substituting guesswork for observation that he was temporarily pulled off teaching duties.

When he had finished his doctorate, Bigelow, with Agassiz's help, persuaded the Bureau of Fisheries to lend him the *Grampus* for a series of short cruises in 1908 to collect medusae in the Gulf of Maine. Bigelow then wrote several papers cataloguing the medusae and siphonophores[*] he had

---

[*]Siphonophores, such as the Portuguese man-of-war, are small, translucent cousins of jellyfish; medusae (which are jellyfish, essentially) and siphonophores are both classified in phylum *Coelenterata* (the coelenterates), which are radially symmetrical and have tentacles and a central body cavity where digestion occurs. Anemones and corals are also coelenterates.

gathered on all his journeys. When his book on the medusae of the eastern Pacific appeared in 1909, "Mr. Agassiz went so far as to tell me that [it] ranked me as a leader among marine biologists, [and] I felt more conceited than I ever have since."[11] That account, still considered a classic, established Bigelow as a first-rate taxonomist, a role he would return to at various times in his life, particularly in its last three decades. In his twenties he also wrote papers on hearing in goldfish, black duck hybrids, and shallow-water deposits in Bermuda.

He also enjoyed his typical cornucopia of strange adventures. He took his eventful Newfoundland honeymoon. He learned to ride hounds after foxes; took long backcountry snowshoe hikes in northern New Hampshire and Maine, paying for bunks in logging camps with magazines he and his friends brought for the isolated, reading-starved loggers; and while returning overland from the 1904 Pacific cruise, accepted his Arizona innkeeper's invitation to hunt prairie lynx from a buckboard—a method that involved traveling across the chaparral at high speed in a bouncing, crashing buckboard while firing revolvers at lynx. Even Bigelow couldn't hit his target under those circumstances.

Some of this merriment cooled in 1910, when Agassiz died. With the loss of his mentor, Bigelow entered a less directed, more uncertain period. He was now a scientist rather than a student, yet he lacked the easy access to fieldwork that Agassiz had provided. He spent more than a year at loose ends, cataloguing and writing up more of the collections at the Museum of Comparative Zoology and finishing his book on siphonophores, but finding no outlet for his desire to do fieldwork.

Then, in 1911, Sir John Murray gave Bigelow a kick in the pants. Dropping by the museum while visiting from Scotland, Murray asked Bigelow what he was up to. The great scientist was apparently chagrined when his old friend's promising protégé told him he wished he could get out on the water but lacked the means. As Bigelow recalls it, the rest of the conversation went as follows:

> "How much is known about the Gulf of Maine?" asked Murray after a pause.
>
> "Practically nothing," replied Bigelow.
>
> "Can you row?"
>
> "Yes, sir."

"Can you borrow a dory?"

"I can do better than that, since I own a sailboat."

"Can't you borrow a deep sea thermometer from the Fish Commission?"

"Yes, sir."

"Can't [you] make some tow nets out of . . . old bobinet window curtains?"

"Yes, sir."

"Don't ask me any more damn foolish questions!"[12]

Murray's prod inspired Bigelow to take up the sort of study he'd been pondering for a year or two as he had learned about the work of the International Council for the Exploration of the Sea (ICES, pronounced "*ice-ease*"), a scientific effort by several European nations to examine the fisheries ecology of the North Sea. Since forming in 1902, the organization had been systematically compiling data on how patterns and variations in current, temperature, depth, and other key factors affected fish populations in the North Sea and other European waters. It struck Bigelow that this scientifically broad but geographically focused method—an approach that would come to be known as "intensive area study," essentially a form of what we now call ecosystem study—would work well in the Gulf of Maine. Of course, ICES was a cooperative venture of several nations employing teams of researchers on numerous vessels over a period of several years. Bigelow was one guy looking for a boat.

By chance—by good scientific luck, as it were—the Gulf of Maine and Georges Bank fisheries had been showing signs of decline that suggested overfishing, and the U.S. Bureau of Fisheries, after years of devoting itself mainly to hatchery work, wanted to conduct some research on the Gulf of Maine's ecology. Bigelow's request for a boat and funding to do an intensive area study landed at the bureau precisely when its directors hoped to conduct such investigations but had no program in place.

Henry Bigelow thus became the bureau's fisheries science program in its most important fishery, and he would remain so for more than a decade. He took on the work even though initially he was more interested in oceanographic work than fishery science. The distinction between oceanography and fishery science is firmer than you might think. At the risk of oversimplifying, oceanography might be said to concern itself with everything about the ocean *except* fish, while fishery science focuses on the question of why and how fish populations expand and contract. As all this happens in the ocean, and as fish dominate the food chain, the two subjects obviously are linked. Yet the fields do not overlap as much as it would seem. As bureaucratic and intellectual endeavors they have usually been pursued as separate disciplines—oceanography more as a pure science and fishery science as a goal-oriented discipline of its own rather than a subdiscipline of oceanography. Alexander Agassiz, for instance, likely never spent ten consecutive minutes wondering how fishing affected fish populations. And Spencer Baird, a man of science who began the Fish Commission (later the Fisheries Bureau) in 1871 under the premise that sound oceanographic work would help manage the fisheries, found that the bureaucratic pressures exerted by the commission's main budgetary rationale—to help the fishing industry—sucked all but token resources away from broad oceanographic research and into the quick fix, which at the time was thought to be hatcheries.

The ICES approach, in asserting that a broad base of oceanographic knowledge could help solve the mystery of fish fluctuations, was an attempt to combine the two disciplines. This notion attracted Bigelow for both intellectual and practical reasons. He liked the idea of examining the big ocean organism that produced fish. And from a mercenary point of view, he saw that an intensive area study of the Gulf of Maine could give him a boat and funding to do significant work in the country's key fishing ground. That it happened to be water off his doorstep sweetened the prospect, for he had teaching duties during the academic year and was starting a family.

4

On July 9, 1912, Henry Bigelow sailed into the familiar, well-traveled *incognitum* of the Gulf of Maine. His first stop was 5 miles outside Gloucester Harbor, where he had the captain swing the bow into the breeze, stopping the boat so that he could test his equipment. The *Grampus,*[*] a retired fishing schooner, had been fitted out hurriedly on a tight budget for that first summer's cruise, and Bigelow wanted to make sure he could effectively use the odd assortment of European oceanographic instruments and standardized and jury-rigged fishing equipment he'd put on board. Gone were the state-of the-art winches and steady platforms he'd worked with on Agassiz's cruises. Because the commission had delivered the boat without any dredging or hauling equipment—no winches or pulleys to bring up the heavy dredges and fishing nets Bigelow brought along—Bigelow installed a gasoline-powered winch he had rigged for his 1908 trip. This device was so anemic he had tied in a second, hand-driven winch to assist it. With this unwieldy rig, he and the crew would have to lower and raise dredges, trawl nets, and plankton nets weighing several hundred pounds— operations made much more difficult by the ship's instability. For the *Grampus,* built for speed rather than stability, rocked terrifically atop the slightest roll when not under way. This made lowering and raising equipment dicey, for once an object was in the water, a sudden roll could pull and snap the line holding it. The ship's rolling also frequently sent Bigelow and others to the rail. (Bigelow's "one physical weakness," a colleague later recalled, "was that on the open ocean he was often seasick."[13] In 1931, when scientists had gathered at the new Woods Hole Oceanographic Institution for its first season of research, Bigelow, who was one of the institution's founders and its first director, gave an inaugural lecture titled "Seas I Have Vomited In"—a subject that covered a lot of water.)

The *Grampus* rocked so much and was so time-consuming to anchor that Bigelow quickly took to performing all but the heaviest operations from a dory anchored near the stern. They took depth soundings the hard way, lowering and raising by hand a 30-pound lead weight on a cod-line (a quarter-inch rope). They installed in the dory a hand winch with 300

---

[*]Grampus is the pilot whale *(Grampus griseus),* also known in New England as the blackfish. This member of the dolphin and whale order, *Cetacea,* resembles a large dolphin, though lacking the dolphin's long snout, or beak. Pilot whales move in groups, and as Henry David Thoreau vividly describes in his book *Cape Cod,* coastal New Englanders, on seeing a group of them off the beach, in times past would hurriedly launch their sailing dories and try to herd them onto the beach—a severe test of sailing skills.

meters of pencil-thick line for dropping and retrieving thermometers, water sampling bottles, and their current gauge.

Bigelow did not scrimp on his thermometers and sampling bottles, however. Both were ingenious devices that had been developed and refined in ICES investigations. The thermometers captured and preserved readings of water temperatures at different depths. The Nansen water-sampling bottles, which locked shut with the most recently scooped water inside when they were stopped in their descent, let Bigelow collect water at specified depths so he could measure its salinity. His measurements of temperature and salinity would, in turn, reveal the water's density (which varies with and is reflected by those combined properties), and together these characteristics would reveal much about where water in a given place came from and how it would flow in relation to waters nearby. As he wrote later, he saw this question of the water's movement as "the most important subject on which the cruise may throw light."[14]

For gathering fish, he had a standard 8-foot beam trawl (a long, cone-shaped net held open by a beam along its bottom edge), and to scoop up fish eggs, larvae, and other plankton,[*] a half-dozen nets of various materials—some silk and some scrim (though none, as Sir John Murray might have been disappointed to hear, from Bigelow's curtains). Some of these could be opened and closed underwater to sample plankton at particular depths. The boat had lab space just off the working deck where Bigelow and his assistants could dissect and identify specimens, bottle and preserve samples, and record observations. (Along with a small Bureau of Fisheries boat crew, Bigelow was variously accompanied by a bureau captain, students from the Museum of Comparative Zoology, his wife, Elizabeth, or, most often, the bureau's William Welsh, a fishery scientist who worked extensively with Bigelow in the Gulf of Maine and would write with him the first edition of the resulting work on fishes.)

---

[*] A rather baggy term, *plankton* (from the Greek *planktos,* meaning "to wander") refers not to a taxonomic category but to the living things that float more or less passively in the ocean. Along with tiny plants and animals, fish eggs and larvae are also plankton, and they compose an important part of plankton hauls and studies. When a fish larva gets big enough to swim—fingernail size—it's no longer plankton. It's a fish.

With all this gear test-hauled and calibrated just outside Gloucester Harbor, Bigelow turned the boat into water he had known half his life, south and west of Cape Ann. The weather was calm and warm, with mild breezes, good for working and sailing. He spent three days in sight of Cape Ann getting the kinks out of the sampling routines. Then, convinced his equipment was seaworthy, he sailed east to sample the deep water of Wilkinson Basin. The water there was so rough that he had to forgo all but the depth, temperature, and salinity measurements from the rocking dory. Finally he headed back inshore to the "proverbially rich" shallower waters of Ipswich Bay, where he began to survey in earnest.

If you look at an ordinary map of the Gulf of Maine, Ipswich Bay, the stretch of water lying between the bump called Cape Ann and the tiny nubbin labeled Cape Neddick, won't exactly jump out at you. Neither will most of the other bays—Massachusetts, Saco, Casco—that gently scallop this western edge of the Gulf. For that matter, you may have trouble finding the Gulf itself the first time you look for it on a big map. It is not an obvious embayment, like the Gulf of Mexico; it's merely a broad, rather shallow indentation into the coast, bracketed by the slim hook of Cape Cod on the west and the fat peninsula of Nova Scotia to the east, but otherwise easy to miss.

An ordinary map, however, does not show the great, thumb-shaped underwater plateau of Georges Bank reaching from Cape Cod almost clear to Nova Scotia. Rising for most of its length to within 50 fathoms of the surface and in many places much closer, Georges encloses the Gulf waters like a gate. (A fathom is 6 feet or just under 2 meters.[*]) The gate doesn't quite close at the eastern end, where the Northeast Channel, more than 100 fathoms deep and 6 miles wide, separates Georges from the shallows and banks off Nova Scotia and provides the only entrance deeper than 50 fathoms. At the western end, what you might call the gate's loose, leaky hinge,

---

[*]The nautical, fishing, and scientific communities vary in their use of terms for depth, with sailors and most fishermen favoring fathoms and most scientific literature preferring meters. Despite meters being the standard internationally and in scientific writing, I use fathoms in this book because it's by far the more common usage in New England among fishermen and is also favored by many scientists when they are not writing for publication.

the Great South Channel, is a shallow trough just shy of 50 fathoms deep that separates the rising plateau of Georges to the east from the wide expanses of Nantucket Shoals to the west. Georges so effectively isolates the inner Gulf's main basin from the Atlantic that the Gulf "may without much exaggeration be considered mediterranean, or landlocked," as Bigelow put it.[15] If you could hover high above and look through the water—or if you look at a good bathymetric chart of the Gulf—you'd see the Gulf quite clearly as an ovoid basin with a barely submerged southern rim separating it from the Atlantic abyss.[16]

Bigelow did not know the full extent of this hydrographic isolation when he started. He knew about Georges, of course, and he knew the Gulf was a good 5 degrees Celsius (9°F) colder than the water south of it. He also knew from centuries of fishermen's observations, as well as from his own experience and the few readings he had taken in 1908, that the Gulf's temperature patterns were ass-backwards, with near-surface temperatures of the water around the shallow rim and over Georges significantly cooler than those over the deeper contours in the Gulf's center. He didn't buy the long-accepted explanation for this, that the Labrador Current, an Arctic flow that slid southwest around Newfoundland, curled around the Nova Scotia peninsula and then circled the rim of the Gulf. That theory didn't jibe with growing evidence from Canadian studies that the Labrador Current petered out before it reached the Northeast Channel. Nor did he buy the idea that cold water welled out of the Atlantic abyss to enter the Gulf by flowing over or around Georges, or the theory that the cold water along the Gulf shore rose from the depths of the Gulf. Wherever they came from, the improbably cold waters of the Gulf's shallows were crucially important, for they provided just the combination of temperature and salinity needed by cod, haddock, flounder, and other fish.

The waters of Ipswich Bay and the other shallows around the Gulf, Bigelow hoped, would reveal where the cold water came from. The first thing Ipswich Bay revealed, however, was lots of fish. This was no surprise, as the Bay had always held lots of fish. Ipswich Bay reaches some 8 or 10 miles across deepening water toward Jeffrey's Ledge, a finger of high bottom that hooks northward from the shallows off Cape Ann. The seabottom between that arc and the Gulf's western shore holds scores of shallows, knobs, basins, and isles fantastically popular as spawning and gathering grounds for fish, porpoises, whales, and seabirds. Fishermen had worked these grounds (the Duck Island Ridges, Boon Island, the Isle of Shoals, the Nipper Ground, Boars Head) for three centuries. So at Bigelow's first sampling station, 4 or 5 miles north of Cape Ann in about 25 fathoms of water, he was not surprised to pull up a full net after dragging the beam trawl a mere half-hour. When he yanked the puckerstring to open the net's cod end, onto the deck spilled—heartily, saltily flopping—almost a hundred flounder; scores of red and white hake, slimy cousins to the cod; a couple of even slimier ocean pout; a dozen skates, cousin to rays; four long, crusty, alligator fish "not much thicker or softer than an iron spike"; and seven large goosefish, a type of anglerfish, also known as the monkfish, whose huge head and horror-movie dentition make it one of the most remarkable, gluttonous, and likeably ugly of the Gulf's many ugly and gluttonous creatures.*

Bigelow found all of this intriguing, especially the red hake. Many of the hake were so ripe to spawn that they oozed, the females leaking sticky clusters of dark eggs, the males streaming pools of milt. The *Grampus* had interrupted an orgy. "This discovery," wrote Bigelow in his report that winter, "is of some interest," because little was then known about where and when this commercially important species spawned.[17] His excitement grew when he and Welsh hoisted up the plankton net at that same station off Cape Ann and found it packed with fertilized eggs similar to the roe found in the red hakes. To make sure they were the same, Bigelow and

---

*In *Fishes of the Gulf of Maine*, Bigelow describes reading "of one [goosefish] that had made a meal of 21 flounders and 1 dogfish, all of marketable size; of half a pailful of cunners, tomcod, and sea bass in another; of 75 herring in a third; and of one that had taken 7 wild ducks at one meal. In fact it is nothing unusual for one to contain at one time a mass of food half as heavy as the fish itself. . . . Fulton . . . found a codling 23 inches long in a British goosefish of only 26 inches, while Field took a winter flounder almost as big as its captor from an American specimen. One that we once gaffed at the surface on Nantucket Shoals contained a haddock 31 inches long weighing 12 pounds, while Captain Atwood long ago described seeing one attempting to swallow another as large as itself." He also writes of finding grebes and other diving fowl in the stomachs of goosefish, the birds evidently captured when they pursued other fish near the bottom. (Bigelow and Schroeder, *Fishes of the Gulf of Maine*, p. 535.)

Welsh performed a little artificial insemination, gently squeezing one of the females over a bucket of saltwater to extrude her roe, then squeezing a male to eject his milt over the eggs. The fertilized eggs from this forced union soon matched those they had pulled up in the plankton net.

Bigelow was pleased to so quickly find something both scientifically interesting and relevant to his bureau sponsors. (In subsequent years, the bureau would catch ripe hake in these waters and use them in its hatchery program.) He and Welsh strained millions more freshly fertilized hake eggs from Ipswich Bay in the days that followed, and vast numbers of the zooplankton *Calanus finmarchicus*. *Calanus*, which at 2.5 millimeters full grown (less than 1/16 of an inch) is a small animal but a big plankton, is easily the Gulf's most abundant zooplankton and a staple of most of its young fish. The reddish brown adults (fishermen call them "cayenne") rise from the Gulf's central basins each spring to drift through the Gulf and over Georges in subsurface clouds; at night, they float atop the water in great rafts. Bigelow found them almost everywhere, often in swarms so large they attracted not only schools of young fish but flocks of petrels; and he learned to anticipate that a haul would bring up abundant *Calanus* if petrels were gathered over water that lacked fish. He also often found spawning hake in the vicinity, suggesting that the hake were spawning there so that the newly hatched larvae would find a ready meal of *Calanus*.

After the big haul of hake, the *Grampus* shuttled out to a basin a couple of miles offshore, did a station there, and then returned west to do another station about a mile outside the 25-fathom contour south of Portsmouth, New Hampshire. They were towing the beam trawl when the ship lurched hard to starboard, the winch screeched and groaned, and the cable snapped and was gone. The trawl net, having snagged on a piece of rough bottom or on one of the hundreds of wrecks that litter the Gulf's shoals, was lost, taking 150 fathoms of wire rope with it. The dredging boom that held the cable over the ship's side was broken as well.

This setback struck on July 18. Sailing to Portsmouth and then to Gloucester for repairs and then waiting for a storm to clear kept Bigelow off the water for four days. On the twenty-second, however, he sailed once again out of Gloucester and zigzagged through the rest of his Ipswich stations, working water off Capes Neddick and Porpoise in Maine that was markedly colder than that just south, then sliding north across Casco Bay to South Harpswell. Casco Bay, up in the Gulf's northwest corner, is lovely, intricate water where the relatively smooth, sandy coastline of the Gulf's western shore makes its transition to the northern shore's rocky, jagged indentations. In summer, its waters are full of striped bass hovering around the mouth of the Kennebec River.

Bigelow wasn't there to hunt stripers; he was there to titrate his water samples at the South Harpswell Marine Laboratory, to see what his salinity tests would reveal. After sailing five stations around Casco Bay to take some bottom samples for the Harpswell lab, he borrowed some lab space at Harpswell and settled into work while Welsh headed east along the coast in the *Grampus* to work more stations.

Titrating seawater is tedious work, and Bigelow would rather have accompanied Welsh. After three weeks of collecting water samples, though, he needed to make sure his bottles were snug. He also wanted to know what they would tell him. To test each sample of seawater, he stirred it, drew a measured amount into a pipette, then dripped water a drop at a time from the pipette into a measured beaker of silver nitrate until the seawater contributed enough chloride to turn the silver nitrate from yellow to orange. Doing this for the multiple samples from the fifteen stations they had done so far took Bigelow five days. To ensure accuracy he titrated each water sample twice, sometimes a third or a fourth time if the results were fuzzy. This occurred fairly often, for he found that if you spent enough time watching yellow turn to orange, the point of transition began to elude you. This was not the sort of work at which Bigelow naturally excelled, but he was determined to do it well. He believed the samples, which were the first reliable saline titrations of any number ever taken from the Gulf, would tell him where the cold water came from and how the waters of the Gulf moved.

The water he tested in South Harpswell seemed to confirm his hunch that the cold coastal water was not coming from the Atlantic abyss or even the depths of the Gulf. It was too watery. Abyssal waters hold about 35 parts of saline per thousand (or "per mille"), or 3.5 percent salt, and the few samples he had taken from the depths of the Gulf were 33 per mille—significantly lower than in the open Atlantic on the south side of Georges. Even more significant, the samples he'd taken of the cool water along the coast ran consistently between 31.5 and 32.5—near the low end of the range for seawater, which runs from around 31 to a maximum of 37, and seemingly far too low to support the idea that the water spinning around the Gulf's big counterclockwise gyre originated in the Labrador Current. He knew it was too early to draw conclusions, but it looked as if the simplest but so far overlooked explanation might be right: The shallower coastal waters were cool—and less saline—because they were flooded every spring and summer by freshwater exiting the many rivers that flowed into the Gulf.

It would take many more samples either to prove or dash this idea, so when he finished his five days of eye-straining titrations, Bigelow reboarded the *Grampus* to cover waters east. He met immediate frustration: After he spent a long day doing a square of five deep stations off Portland, a storm drove him back to port for five days. On August 13, feeling the summer closing (he had teaching duties awaiting on the first of September), he stood the *Grampus* due east toward German Bank, a small but important fishing shallow off the tip of Nova Scotia, at the far eastern end of the Gulf. He must have caught a strong westerly, for the *Grampus* made this 200-mile run, crossing the entire Gulf, in just over 36 hours despite stopping en route for four time-consuming deep-water stations. They made stations 29 and 30 on German Bank "on the evening of August 14th in thick fog," then another station the next day a bit north in fog so dense he could not ascertain their exact position.[18] They were off the mouth of the Bay of Fundy now, the great long inlet between Nova Scotia and New Brunswick, and heading northwest toward the far eastern reaches of the Maine coast. On Great Manan Bank, they came upon a fishing boat "making a good fare of cod." That night, the fog lifting, they worked through the night, taking a rich surface haul of *Calanus* about three in the morning under the dim light of a crescent moon.

As they neared the northern shore they found something unexpected: The plankton bloom they had been sailing through almost everywhere suddenly disappeared. Swinging back and forth along the coast, Bigelow found that the *Calanus* dropped off to nothing every time he crossed the 100-fathom contour to shoaler water. This coastal band of barren water proved small, however, as they again hit rich blooms of plankton when they headed west along the coast. Their route straightened now, with long intervals between stations, for despite the beauty of the wild, jaggy coast and their own need for data, they had to run home. The schedule was pushed tighter yet when weather again drove them into Casco Bay for several days and harassed them further on their course south toward Gloucester. They pulled into Gloucester on August 31. They had spent most of seven weeks at sea and collected hundreds of plankton, fish, and water samples at forty-six complicated stations. Their last station was just outside Gloucester Harbor, where they stopped once more at the end of their voyage to take some readings they could compare to those they had taken earlier.

The Gulf's days as a *mare incognitum* had ended. Over the winter and spring, as Welsh took short bimonthly cruises to track changes in temperature, current, and salinity, Bigelow composed a one hundred-page report of the summer's cruise, complete with numerous tables and a chart showing the journey's stations, for the museum's bulletin. Though he noted that much more investigation was needed and that he had not yet worked the vital water over Georges Bank, Bigelow wrote that it seemed clear that the water along the Gulf of Maine coast was cold and dilute because of freshwater pouring in from rivers and that "if any general conclusion can be drawn from the scanty oceanographic data yet available, it is that the Gulf of Maine owes its low temperature and salinity largely to local causes; i.e., to its geographic position and partial isolation by the sill formed by Georges Bank."[19]

This conclusion would hold up and be refined over the years as Bigelow, Welsh, and others continued to work the Gulf. Bigelow sailed extensively almost every summer for the next fifteen years, extending his sampling stations onto Georges Bank in 1913 and 1914 and then methodically expanding and repeating his scattered web of stations. As he streamlined his collection procedures and gathered a more solid base of oceanographic data, he had more time to collect fish. He and Welsh also queried fish dealers and port agents and examined landings records up and down the coast to see what was being caught where. They grilled knowledgeable fishermen about where the fish were and how thick they were running. Bigelow was adamant that you had to talk to fishermen to know fish. Once, in 1920, when he had broken his arm in some accident and Welsh was unavailable for an early cruise to look for spawning haddock, Bigelow was so concerned that the bureau employee assigned to the mission, a Mr. Rankin, would simply wander around looking for fish that he relayed through the bureau's deputy commissioner, H. L. Moore, specific instructions for finding the haddock: "Select two localities where Haddock are spawning and where at the same time the bottom is sufficiently smooth for trawling. One of the stations should be near the Isle of Shoals and the other on Stellwagen Bank or off Cape Cod, but the precise spots can only be determined after consulting the local fishermen."[20] Worried that Rankin would blow it anyway, Bigelow went along after all, despite his cast. As the rocking *Grampus* routinely made necessary the "sailor's grip" (work with one hand, hang on with the other) we can only imagine the aggravations Bigelow endured on that cruise. When it ended he wrote Moore, "We . . . successfully carried out all the stations. Personally I am glad it is over as a broken arm is an inconvenient thing at sea."

By this time, Bigelow was forty, and after eight summers he was sailing more for the data than the pleasure of being on the water. The pleasure, after all, he could experience with less discomfort on his own sailboat, on which he could choose his crew and which he did not have to stop for long sampling stations that forced him to suffer the *Grampus*'s merciless rocking. Bigelow wasn't a complainer. But there must have been moments while he was away from his young family those long weeks at sea—perhaps as he clung to the rail contributing plankton of his own to the Gulf of Maine—when the project seemed endless and the *Grampus* a curse. He would have had to call on all the discipline he had learned working with Agassiz, pushing himself to stay at each nauseous station until he had it all—the multiple temperatures, the duplicate water samples at different depths, the numerous plankton hauls—and to resist the temptation to say, That's enough, it's too rough, let's make sail. Yet he did the work so exhaustively that it took decades to substantially supplement it.[*]

---

[*]It wasn't until the 1970s, when proposals for oil drilling on Georges Bank spurred a huge research program, that scientists added enough to Bigelow's findings about the Gulf's physical oceanography to displace his 1927 volume as the standard text, and those and subsequent investigations have merely elaborated on rather than revised his research. As for his work on fishes, his 1953 *Fishes of the Gulf of Maine,* a revision of his 1925 guide of the same name, remains the standard reference on the subject.

➤◄

5

By the early 1920s, with the Fisheries Bureau making noises about retiring the *Grampus,* Bigelow turned to the task of writing up the results of his long investigation. He was not exactly facing a pile of undigested material, as he'd been compiling and summarizing the results of his cruises all along, sometimes for publication, and he'd seen distinct patterns emerging.

The most developed pattern was that of the Gulf's currents, which he depicted in a map that has been endlessly studied, cited, reproduced, and relied on ever since.* The map fills with admirable detail and accuracy a canvas that had been virtually blank when he started.

The map presents the short version. Bigelow offers the long version (along with the map) in his *Physical Oceanography of the Gulf of Maine—* 516 pages and 207 figures occupying an entire volume of the *Bulletin of the U.S. Bureau of Fisheries.*[21] This extraordinary tome draws on Bigelow's vast fieldwork and the latest contemporary mathematical models of how temperature, density gradients, and the earth's rotation affect current to build a picture of the Gulf's complex hydrography. The picture it describes is so dynamic and three-dimensional that it seems architectural.

Noting that "the physical laws involved are . . . the simplest, embodied in the old adage 'water runs downhill,'" Bigelow describes how variations in the water's density, along with the massive inflow from rivers, create in the Gulf of Maine a slight but vital downhill slope of the water's surface from the shoreline out toward the Gulf's opening.[22] All else being equal, the slope would cause the Gulf's water to slide out evenly over the lip formed by Georges Bank. But, he explains, citing a recent explication by his friend A. G. Huntsman of Canada, this even flow is warped by the Coriolis effect caused by the earth's rotation, which in the Northern Hemisphere tends to create counterclockwise gyres in basins and clockwise gyres around islands and banks.[23] Intensifying the Coriolis effect in the Gulf are

---

*The essence of the currents map is overlain on the map in the front of this book.

tides, variations in the density of water, and the Gulf's basinlike shape. The Gulf's counterclockwise gyre thus easily resists corruption from the strong tides that pull water in and out of the Gulf over Georges Bank. (In fact, that tide is made circular by the gyre.) The gyre also pulls some water (and a small but revealing number of tropical immigrant plankton) in through the Northeast Channel and spins it around the Gulf rim, and it sends the Gulf's surplus of water (for enough freshwater flows and falls into the Gulf each year to raise its level 3 feet) gushing out the Great South Channel. The Gulf's counterclockwise gyre also bumps against and feeds water to the clockwise Georges Bank gyre, supplying it with the abundance of plankton that makes Georges such a spectacular fish nursery.

Bigelow's close study of densities, temperatures, and current, fed into contemporary hydrodynamic formulas, also began to explain the remarkable insulation that the Gulf of Maine and Georges Bank preserve from the much warmer Gulf Stream water that bumps against Georges' southern edge. Here, where the cold, relatively dilute Gulf water pushes against the warmer, more saline ocean water, contour lines representing changes in nearly any measurement you can take—salinity, temperature, density, current—press snugly against each other like the isobars defining a dense cold front or the contour lines of a cliff. Along this compact density and temperature gradient Bigelow found a confused column of water, a "cold wall," that effectively separates the warm, eastward-flowing water of the Gulf Stream from the colder, denser, westward-flowing water on Georges' southern flank. With the cold wall as a buffer between them, Bigelow concluded, the two masses of water slide against each other somewhat as oil slides against water. Bigelow also posited a narrow trough of low water over this cold wall—an actual depression in the ocean's surface over the seam between the Gulf water and the Atlantic water—that both responded to and strengthened the separation. Decades later, extensive computerized studies employing satellites, lasers, numerous current meters, and other high technology have proved Bigelow's description to be essentially correct. With his uncanny knack for grasping the heart of a problem, Bigelow had defined the basic dynamics of a phenomenon so complex that teams of bright scientists with supercomputers were still struggling to understand it seventy-five years later.

These oceanographic findings helped explain much about the Gulf's ecology. The maps of the Gulf's and Georges' currents, for instance, explain how the gazillions of *Calanus* and other zooplankton that rise out of the Gulf's deep basins each spring and summer happen to float in great clouds over the shoals of the Gulf and Georges just as hundreds of millions of cod, haddock, pollock, and other larvae in those shallows are turning into hungry fishlings. Researchers have produced spellbinding simulations of these plankton blooms viewable on home computers: From the Gulf's dark basins ascend clusters of specks that spread, their colors changing as they age, into the Gulf's counterclockwise coil.[24] All through May, June, July, and August, the gyre pumps them onto the shallows, sending some swinging against the western shore and others into the Great South Channel, where Georges' clockwise gyre grabs them and pulls them over and around its great mound. This animation uses complex mathematical models to visually convey data drawn from satellite-tracked drift particles, but they are essentially confirming and elaborating on Bigelow's findings.

The fascination of watching these simulations, entrancing as they are, pales next to that of reading Bigelow. Though he can grow opaque when explaining current dynamics, Bigelow usually writes with concision, vigor, and color, integrating effortlessly and with flawless judgment the "hard data" of scientific studies and the more vivid anecdotal observations from his own and others' time on the water.

His art reaches its height in *Fishes of the Gulf of Maine*, which he wrote first with Welsh in 1924 and then with another longtime collaborator, William "Captain Bill" Schroeder, in a 1953 edition that incorporated the three decades of intervening findings, data, and anecdotes. Bigelow wasn't even supposed to be the lead author on the first edition of *Fishes*. Welsh was to write it while Bigelow composed the volumes on physical oceanography and plankton. But when Welsh died, in 1921, while compiling the notes, Bigelow had to take up the task. The sharp-eyed, well-traveled Welsh's knowledge and perspective were essential, as were Schroeder's. Yet both editions are clearly Bigelow's work, imbued with his distinctive voice and the particular pleasure he took in grasping the essential patterns of the ocean's dynamics, whether they involved currents, migration, or a fish's feeding habits, and the variations of those patterns.

The first time I happened upon *Fishes,* in the wet lab of the National Marine Fisheries Service research vessel *Delaware II* between trawl hauls over Georges Bank, the great book—a ream thick, encased in a worn, brick-red library binding—fell open to this entry on the great bluefin tuna:

> This is the largest Gulf of Maine fish, except for some sharks; a length of 14 feet or more, and a weight of 1,600 pounds being rumored, with fish of 1,000 pounds not rare. The heaviest Rhode Island fish on record, taken about 1913, weighed 1,225 pounds, while 4 or 5 fish have been brought into Boston that weighed approximately 1,200 pounds each, and one in 1924 that is said to have reached 1,300 pounds; and Sella mentions a "fairly well authenticated instance" of one caught 60 to 70 years ago off Narragansett Pier, R.I., that weighed in the neighborhood of 1,500 pounds, was divided among the various hotels, and fed 1,000 people. . . . .
>
> The tuna is a strong, swift fish and an oceanic wanderer like all its tribe. . . .
>
> When tuna are at the surface, as they often are, they are proverbial for their habit of jumping, either singly or in schools; they may do this when swimming about, or harrying smaller fishes, or less often, when traveling in a definite direction, when all that are jumping go in the same direction.
>
> Frank Mather, for instance, reports seeing a school of 200-pounders jumping in union 2 or 3 feet clear of the water. When large tuna jump, they sometimes fall flat, making a great splash, but they reenter the water a little head-first as a rule, though they do not make as complete and graceful an arc in the air as the various oceanic kinds of porpoises usually do. When schools at the surface are not jumping, they often splash a good deal and are conspicuous then. We remember, for instance, sighting a large school so employed, off the Cohasset shore at a distance of about 3 miles, on one occasion.[25]

After describing the tuna's rather catholic tastes in food, Bigelow notes the one danger, other than humans with hooks or harpoons, that confronts this swiftest and most magnificent of Atlantic fish: "Tuna have no serious

enemies in the Gulf of Maine, but killer whales take toll of them in Newfoundland waters where, writes Wulff, 'one or more times annually, usually
in September, orcas will ravage the tuna schools in the bays they frequent
most.'"[26]

Clearly this was no ordinary field guide. Flipping to the front of the
book, I dove into Bigelow's entry on the hagfish, also known as the slime
eel. This creature occupies the opposite end of the scale of grace and
grandeur from tuna. *Myxine glutinosa,* an eyeless, lampreylike fish 1 to 2
feet long, with a finless, tubular, wormlike body, scaleless skin, lipless
mouth, and rasp-sharp tongue, is a sort of nightmare of the sea. This scav-
enger lives on the cold bottom and feeds

> chiefly on fish, dead or disabled, though no doubt any other
> carrion would serve it equally well. It is best known for its trou-
> blesome habit of boring into the body cavities of hooked or
> gilled fishes [that is, of fishes caught on longlines or gillnets left
> in the water to be retrieved], eating out the intestines first and
> then the meat, and leaving nothing but a bag of skin and bones,
> inside of which the hag itself is often hauled aboard, or cling-
> ing to the sides of a fish it has just attacked. In fact, it is only in
> this way, or entangled on lines, that hags ordinarily are taken or
> seen.
>
> Being worthless itself, the hag is an unmitigated nuisance,
> and a particularly loathsome one owing to its habit of pouring
> out slime from its mucous sacs in quantity out of all propor-
> tion to its size. One hag, it is said, can easily fill a 2-gallon
> bucket, nor do we think this any exaggeration.[27]

Equally vivid is his account of the sad fate the Gulf of Maine offers the sunfish *Mola mola.* The sunfish is a huge, freakishly beautiful warm-water fish
that occasionally wanders off the Gulf Stream into the Gulf of Maine,
where the cold waters leave it "chilled into partial insensibility."

> They float awash on the surface, feebly fanning with one or the
> other fin, the personification of helplessness. Usually they pay no
> attention to the approach of a boat, but we have seen one come
> to life with surprising suddenness and sound swiftly, sculling
> with strong fin strokes, just before we came within harpoon
> range. When one is struck it struggles and thrashes vigorously
> while the tackle is being slung to hoist it aboard, suggesting that

they are far more active in their native haunts than their feeble movements in fatally cold surroundings might suggest. . . .

The sunfish is described as glowing luminescent at night in the water. We cannot verify this first hand. But we can bear witness that it grunts or groans when hauled out of the water.[28]

It is rare to find a finer observer or better teacher of natural history or to witness a more successful, vivid integration of disparate sources of information.

Welsh's death, and the pressure both on and from the Bureau of Fisheries to produce some clear, fishery-related product from its decade of research, compelled Bigelow to write *Fishes* before the works on plankton (1926) and physical oceanography (1927). This prevented him from folding his full conclusions on those subjects into *Fishes*. Yet he knew the currents, salinities, tides, flows, and temperatures; he knew where the plankton bloomed and where they drifted; and he could and did weave that knowledge, along with his thousands of pages of reading, the hundreds of conversations he and Welsh had with fishermen, and his own countless hours on the water, into the field guide.

The richest entries give not only a vivid picture of a fish's character and existence but of its relationship to humans, and Bigelow always extends to its human pursuers the same sympathy and respect he has for the fish. This respect isn't expressed overtly but by Bigelow's matter-of-fact acceptance of the fishermen's careful observations as valid and relevant and by his seamless inclusion of fishermen's observations with those of scientists. His long account of the haddock, for instance—a commercially valuable species, cousin to the cod, which was a prime focus of fishing efforts and of an extended tagging program Bigelow participated in during the 1920s—includes a judicious synthesis of information regarding the fish's whereabouts and movements:

> Our fishermen have long realized that the larger haddock, like the larger cod, are so constantly on the move in search of food that the fishing may be poor tomorrow where it was good today, or vice versa. An analysis of the catches that we made on Nantucket Shoals during the tagging campaigns of the U.S. Bureau of Fisheries, 1923–1931, shows that considerable changes took place in the abundance of fish within periods of a few days or weeks at the spots fished. . . .

> Fishermen have long been aware that the haddock vary widely in abundance from year to year and over periods of years, on one ground or another, independent of any effects the fishery may have had on the numbers of fish. It has been amply proved by investigations . . . that the fluctuations result chiefly from differences from year to year in the number of young that survive and take to the bottom on the grounds in question.[29]

Bigelow's regard for fishermen did not blind him to the consequences of their escalating efforts and technology. He describes the halving of the haddock population during the early 1930s, when demand for haddock from the growing frozen fish market sent fishermen looking for them with newly developed, vastly more efficient trawl nets. After rising for a time in the late 1920s, catches declined sharply after 1930, a phenomenon Bigelow concludes "is only too clear evidence of overfishing."[30]

He offers, too, the even grimmer story of the halibut, which had been virtually wiped out the previous century by longliners—boats that caught the fish with long strings of baited hooks they would leave at the bottom and return for later that day. Even with this gear, far less efficient than trawl nets, nineteenth-century New England and Nova Scotian fishermen hunted halibut so effectively that in 1924, Bigelow noted, "The history of the halibut in the Gulf of Maine, like that of the salmon, must be written largely in the past tense."[31]

Halibut are the largest flatfish—huge flounders, essentially, with the flounders' distinctive flattened body, pale underside, and rather grotesque migration of both eyes to its "upside" (that is, the side that will face up when the fish lies on the bottom).[*] But while most flounders top out at the

---

[*]The migration of the eyes to one side, and the rest of any flatfish's side-to-side differences, occurs during the larval stage. A freshly hatched halibut larva, for instance—all one-half inch of it—has a fairly normal, symmetrical, "fish" body with identical sides and with eyes on both sides of its head. As the larva grows, its left side becomes pale, its right side dark and mottled, and the left eye migrates slowly over the top and then to the right side of the head. The mouth, however, stays put, that is, perpendicular to the axis of its flattening, widening body. Flatfish species are either "right-handed" or "left-handed." A right-handed fish will face to your right when placed on a surface with its dorsal (or topmost) fin farthest from you; a left-handed flounder will face left. Most Gulf of Maine flatfish are right-handed.

size of a dinner plate or a trash-can lid, full-grown halibut run 4 to 5 feet long and weigh 200 to 300 pounds, and a few giants of old reached 6 feet long, 2 feet wide, and 600 or even 700 pounds—less a dinner plate than a dinner table, and a hefty one at that. Like most flatfish, halibut get many of their meals by lying unnoticed on the bottom and then rushing up to gobble passersby. They gobble indiscriminately. Bigelow reports halibut eating cod, haddock, cusk, rosefish, sculpins, silver hake, herring, capelin, other flounders ("these seem to be their main dependence"), skates, wolffish, grenadiers, mackerel, crabs, lobsters, clams, mussels, "and even sea birds," citing accounts of halibut that ate a razor-billed auk, a dovekie, and a gull. Fishermen told him of others that ate the heads and backbones of cod thrown overboard "and a variety of indigestible objects such as pieces of wood or iron and even fragments of drift ice." A hungry fish. On one occasion, writes Bigelow, scientists saw a halibut of about 70 pounds at the surface trying to kill a small cod by beating it with its tail. The scientists rowed a dory out and gaffed both halibut and cod.

Until the early 1800s, halibut were abundant throughout northern New England and Canadian waters. They were considered poor fare and were left alone. Cod fishermen thought them a nuisance. Then, in the 1830s, demand for them developed in Boston and New York, and fishermen started fishing them hard. The first years produced spectacular catches: 400 fish in two days by four men with hand lines; 250 fish in three hours; a boat completely loaded in two days; another vessel landing 20,000 pounds in one day. By 1850, the nearer parts of the Gulf and Georges were so empty of halibut that it no longer paid to go to those waters to pursue them. Big boats hunted them at the far edges of the Gulf, at Browns Bank and the Northeast Channel, and on Georges' southeast slope. After they fished halibut into scarcity in those areas, fishermen followed them to the Grand Banks off Nova Scotia, then to the shoals west of Greenland, to the northernmost reaches of the Gulf of St. Lawrence, and eventually to Iceland. By 1890, practically all the halibut being landed at Gloucester was coming from Iceland.

By the time Bigelow and Welsh wrote this sad account in 1924, no one sought halibut but a few fishermen fond of the great beasts: "We have enjoyed the acquaintance of several fishermen, especially interested in halibut, who treasure to themselves a hard-gained knowledge of particular spots, not too far offshore, where they are likely to catch one, in a day's pleasure." Otherwise, forget it. The halibut had become so rare it was not worth fishing for—what we now call "commercially extinct." Trawlers still caught halibut in their searches for other groundfish, but even with the greater efficiency of the trawl gear, landings by the mid-twentieth century were a tenth what they had been a century earlier.

Few of the stories in *Fishes* are so sobering. More commonly the entries convey a teeming, tumbling, almost boisterous vitality—those leaping tuna and the orcas taking toll of them; goosefish grabbing grebes and widgeons; mummichogs burrowing into mud when the outgoing tide leaves them stranded; seahorses "dwell[ing] chiefly among eelgrass and seaweed, where they cling with their prehensile tails, monkeylike, to some stalk." At one point Bigelow tells of "an army of silver hake harrying a school of small herring on a shelving beach at Cohasset, Mass. We half-filled our canoe with pursuers and pursued, with our bare hands."

This last anecdote is from Bigelow's youth at Cohasset, and the passage is typical in its graceful blending of first-person experience with both the structured and casual observations of other scientists and fishermen and with what might be called "common knowledge."

Reading these entries for the first time while on the *Delaware II* on Georges Bank in early 1998, as I did, was both a heartening and a discouraging experience. We were conducting the National Marine Fisheries Service spring groundfish survey, dropping the net for half-hour tows at randomly generated coordinates to see what fish were down there, and the book eased tremendously the job I had as a volunteer scientist in identifying what came up in the nets. More to the point, the richness of Bigelow's accounts gave detail and context and lives to the fish that came flopping into the checker. It gave a sharper sense of what was going on under the sea's rolling opacity.

It was also a pleasure to read of this scientist so thoroughly enmeshed in the lives of the fish and the people who fished for them. His flawlessly judicious use of what today is deemed "anecdotal evidence" (the stories fishermen tell and the information those stories contain) conveyed an inspiring faith: While poorly observed or distorted information from fishermen was worse than worthless, the best and most astutely observed information was priceless—a polarity that holds for the observations of scientists, as well, but which few scientists these days have the courage to apply to information from "nonscientific" sources.

In this I found hope that the corrosive hostility that had developed between fishermen and scientists since Bigelow's time, which had utterly crippled their ability to work together to reverse the disastrous crash in the Gulf's fish and its fishing industry, could perhaps be remedied—that the gap between fishermen and scientists and their perspectives was bridgeable. Hell, if Bigelow—a silver-spooned, crusty, Boston patrician with three degrees from Harvard and a vicious intolerance of shoddy science could bridge the chasm, maybe today's scientists and fishermen could, too.

On the other hand, the business about the halibut was depressing. I hadn't heard about the halibut before, and it was not pretty to read of now. We were on some of the most productive sites of Georges Bank looking for cod—searching the heart of the world's most fertile fishery for the fish that formed its blood, a creature that exerted a pull on fishermen and scientists alike far beyond anything explainable by its monetary value or even by its beauty, which frankly takes a while to see—and we weren't catching a fucking thing.

PART II

I

"We'll go slow these first few," said John Galbraith to the volunteer-scientists-in-training gathered around the fish checker on the rear deck of the *Delaware II*. "You'll get the hang of this sooner than you think."

We were a half-day into the National Marine Fisheries Service's annual twelve-day spring New England groundfish survey, headed for Georges Bank to count fish. Above us, gulls keened as they vied for our scraps. On either side, the sea rolled gently. Behind us gaped the opening to the 12-foot-wide, 30-degree ramp up which the trawl net had been hauled—and down which, it seemed to me, anyone might easily slide into the water roiling at the stern. Only a thin chain guarded the ramp; it was easy to imagine a lurch of the boat, a slip on the wet deck, and good-bye volunteer scientist. Farther astern, barely lost from sight over the earth's curve, was the trammeled but still lovely shore of New England. There waited scientists, regulators, and fishermen—the fishermen rolling their eyes and shaking their heads, discounting our findings before we found them—to see what we would catch.

We volunteer scientists, meanwhile, were wondering what we had just caught. Before us in the checker—a 4-by-8–foot wooden box 10 inches deep and mounted on a waist-high frame—lay a discouragingly diverse pile of fish that had wetly sloughed from the net a moment before. The collection seemed calculated to flummox taxonomic novices. Some were long, slippery, eel-like fish. Some resembled manta rays. Some I knew to be flounders, though not what kind. There were toadlike fish with spikes poking from their skulls. Fat, pink, cheerful-looking fish lay jumbled among serious-looking steel-blue missiles. A few lobsters glared up from amid the chaotic pile, claws a-ready. Anchoring the whole arrangement, half-buried beneath the rest and sadly inert, were a few big greenish fish I was fairly certain were cod.

I was not alone in feeling slightly lost. NMFS's Northeast Fisheries Science Center (NEFSC), which runs the assessment programs for the New England region, grants the title of volunteer scientist on its research cruises more in recognition of volunteerism than science. NMFS's reliance on volunteers exercised a principle sworn to by Henry Bigelow when he was director of Woods Hole Oceanographic Institution in the 1930s; Bigelow believed most intelligent, motivated laypeople could do first-rate oceanography and fishery science fieldwork, and many of his trainees proved him right. For the Northeast Fisheries Science Center, using volunteers made it possible to staff the center's almost continuous series of ocean trips— dozens of trips a year—where two shifts of six scientists work twelve hours a day for twelve days straight. On a given cruise, volunteers composed about half the scientific crew, the rest coming from NMFS staffers corralled into duty. The present crew was more or less typical. Besides myself, gathered around the checker for our taxonomic field seminar were two other first-time volunteers; another volunteer who was on his second consecutive cruise; a fairly new NMFS employee on her first cruise; and a NMFS economist on his fifth. Say "port" in this crowd and you could see people trying to remember whether that meant left or right. In the face of our inexperience, John Galbraith's professed confidence that we would get the hang of this seemed both charitable and brave.

This was not the Harvard economist, nor was he related to him, though this John Galbraith was, like the economist, six and a half feet tall, and he took to water and fish the way the more famous Galbraith took to economics and politics. His interest was professional, personal, recreational, aesthetic, and otherwise. He liked studying fish, and he loved catching them. As the more famous Galbraith was a Keynesian, so this Galbraith was what I call a McGuanist, after the novelist and angler Thomas McGuane, who once opened a travel-magazine story with the sentence, "I fish all the time when I'm at home; so when I get a chance to go on a vacation, I make sure I get in plenty of fishing." Galbraith was now on the fifth of six consecutive twelve-day cruises. He had begun this stretch by counting frozen scallops on Georges in February, and he was now close to ending it by venturing out to Georges again in April. He had constructed this grueling schedule so that he could return home in late April and be home all summer to fish. As he had since childhood, he would fish for the stripers that flowed around Cape Cod all summer, for the flukes that slid inshore as the waters warmed, and for the bonito and false albacore that arrived in July and August. Bonito and false albacore are small tuna, and Galbraith loved to catch tuna. He once took a rod on a summer NMFS assessment cruise that cut southward through tuna water, and from the broad deck of the *Albatross IV*—a 187-foot boat capable of catching several tons of fish at a time—he cast for tuna. He took such a hazing that he never brought the gear again. "Besides," he told me, "as chief scientist you're really supposed to be on duty all the time, and I realized it just didn't look right."

As the person in charge of stocking and staffing the assessment cruises, however, Galbraith could schedule his sea time in the fishing off-season so that when he was ashore—or rather, not *way* out at sea—he could get in plenty of fishing. "He was always that way," his mother told me when I asked her about it. "*Has* to be near the water." Once, when he was three years old, she said, she could not find him anywhere in or around their shoreside home on the Cape. As a hastily recruited search party fanned out looking for him, she grew sickeningly convinced that he had wandered the hundred yards to the beach and drowned. Finally, a neighbor found him a half-mile down the shore. He was wading into a small cove, dragging a bucket in one hand and a net in the other. He was, he announced, going "cwabbing."

Galbraith had been combing the waters around the Cape ever since, constantly expanding his range of known water, first exploring Buzzard's Bay in small skiffs, then the Nantucket Shoals in a Boston Whaler and more lately a 21-footer, and for the past eight years, all of the eastern seaboard, Georges Bank, and the Gulf of Maine as a NMFS employee. Though only in his early thirties, he had already logged almost a thousand days on NMFS cruises, making him one of the more sea-seasoned employees. Those sea days, along with a sharp eye, graduate work in fisheries biology, some taxonomy courses, and his countless hours exploring the water by hook, had made him one of the best field taxonomists in the service. On a later cruise I watched as another seasoned NMFS biologist, Jay Burnett, and three colleagues spent a good quarter-hour applying their more than fifteen-hundred days of taxonomic experience (and thorough searches of Bigelow's *Fishes* and the NMFS field guide) to the problem of classifying a small specimen whose flat body and scooped forehead declared it a filefish, but whose youth obscured its full lineage. "Save it for John," Burnett finally said, and someone put the little fish in a Ziplock and penned Galbraith's name on it. When we got back to shore, Galbraith identified it after maybe five seconds.

"The guy is good," said Burnett.

Though his modesty makes it unlikely he would, Galbraith can also claim humor, humility, and great patience among his virtues. These make him an excellent trainer of volunteer scientists.

"Let's start with an easy one," he said. He reached into the bin, grabbed a tail, and extracted one of my tentatively, silently identified cod. The fish arched slowly, then relaxed, either resigned or dead.

"Any clues?" Galbraith asked.

"Cod?" one of the other volunteers offered.

"Well, it does look like a cod at first," said Galbraith, smiling and pushing up his glasses. "Sort of. And it's related. It's a haddock. Both from the *gadid* family." He reached across the pile, grabbed another tail and pulled, with a sort of slipping, sucking sound, a second, darker fish next to the first. "Here's a cod. Easier to tell the difference when they're side by side. You see, the cod has these spots, kind of like a trout's? No spots on the haddock. And you see how the cod's lateral line," he said, tracing with his finger a thin line along the cod's side, "takes this nice upward curve where it goes over the fat part of the fish? And it's a pale line on the cod? On the haddock this line is black, and it takes a much smaller upward curve. It's almost straight. Also—this is the real giveaway—the haddock has this dark spot right over the pectoral fin. It's called 'the thumbprint of St. Peter.' Something to do with Peter culling this fish out at some point, I can't remember the story. At any rate, the thumbprint gives this fish away as a haddock. And this is typical. You've got these two fish that look a lot alike at first; they're both gadids; but if you work with them a little bit, you see they're really very different. By tomorrow you won't even think about all these marks. You'll just look and you'll know: cod, haddock. You'll just throw 'em in buckets."

Which he did, lifting the two fish by their tails and dropping them into a couple of orange laundry baskets behind him.

From there he went to skates. These "disc-like" fish, as Bigelow describes them, cousins to the rays and "thin as a shingle toward their outer edges," proved trickier.[1] They range from the size of large pancakes to jumbo pizzas. The large, sharp knobs along the thorny skate's spine make it an easy call, as do the leopard spots of the rarely seen leopard skate.

But the females of the two most common skates, the little and the winter (the latter also known as the big skate), are easy to confuse, since juvenile female big skates, not being very big yet, closely resemble adult female little skates. To discern the difference you must examine their teeth, of which you'll find more in the big skates. Males of these two species are easy, for compared to juvenile big skates the adult little skates sport much bigger claspers—long, fragile, pink, fingerlike appendages that flank the tail and grip the female during mating. As Galbraith put it, "Big claspers, little skate. Easy."

Flatfish, or flounders, also perplex. Some you can peg easily by color, markings, or mouth size. More typically, however, you must detect subtle variations of shading and spotting. It doesn't help that each flounder has several common names that often overlap with other flounders' names. Thus, the American plaice is also a dab, though not a mud dab, which is another name for a winter flounder, which is also called a sole, not to be confused with a gray sole, however, which is a witch flounder, which flounder is also a Craig fluke but not a plain fluke, which is a summer flounder. It made you damned glad to come across a yellowtail flounder, which has a yellow tail, a yellow underside, and no other common names, and happier still to pick up a four-spotted flounder, which has those four spots and is what it is and nothing else.

Worst are hakes. Some hake species so closely resemble one another that you can identify them only by pulling open their gill covers and counting the gill rakers, the tooth-like bony structures arranged in a row within the gill to keep food and other solids from entering the gill's respiratory membranes. It took me two full cruises to tell a white from a red, and even then I sometimes mixed them up. Blueback herring and alewives are almost as bad. To identify one, you have to break it open along the ventral seam with a little snapping motion of your hands, like breaking a stick, so you can check the color of the lining of its abdominal cavity. Dark blue, it's a blueback; pale whitish pink, alewife. Best forget that on Cape Cod some people call alewives herring.

Thus we learned, with Galbraith's help and between-haul dips into Bigelow and Schroeder and the NMFS field guide, to solve the first puzzle presented by these fish: which was which. This simple task yielded immense pleasure. Part of it was the drawer-organizing satisfaction of ordering a chaotic pile into neat buckets and then even neater marks on taxonomic charts—all while standing on a rolling boat that fought to keep its own order atop a tumbling sea. This apparent ability to maintain order amid nature's chaos provided the same reassurance one found in the secure niches and drawers that kept our books and toiletries from flying about our cabins and (in rougher seas) the bars on our bunks (sissy bars, the sea-toughened boat crew called them) meant to keep us from being tossed to the floor.

Yet the greatest pleasure of this taxonomic chore lay in encountering the unpredictable, seemingly playful richness of the creatures we fetched and of their wonderful names, which so often evoked and echoed their particular characters, forms, and colors: the chesty-bodied, stumpy-winged sea robin; the streamlined, staccato-striped, muscle-hard mackerel; the supple, subtly colored loligo squid. Then there were the fish you *might* pull up, alluring promises from the field guides: puffers, balloonfish, cow-fish. Boarfish, frogfish, lump suckers. Tonguefish. The spadefish, shaped like the garden implement. Swelltoads and stingerees. The hog choker. Pigfish and white grunts. The pearly razorfish, the freckled soapfish, the slippery dick. The sea mouse. These were not names someone would invent at an exam table on shore. They were sea names. They asserted both the vivid essence of the creatures they described and the salty attentiveness the fish's original namers brought to the sea.

2

Naming, of course, does more than express understanding; it makes it possible in the first place. Thus, our voyage. We were headed to Georges to name things, then count them, so that the Northeast Fisheries Science Center could understand how many fish were out there, who lived where, who was eating whom, and, with luck, how it all fit together. In technical terms, our fish-counting work would produce an "index of relative abundance" of the various species we came across (an indicator of how many fish were out there compared to previous years), which the Science Center's assessment team would use as a major component in its complex formula for producing its stock assessments. As everyone on board knew, the value of both the survey cruises and the stock assessments they helped produce had become a hot subject in the debate over how to manage New England's grievously troubled fishery. On one hand, the Science Center's spokespeople and assessment scientists liked to note that every time the crew of the *Delaware* or its sister ship, the *Albatross IV*, pulled a fish out of the water and weighed, measured, and tallied it, it added to "one of the longest-running and most complete wildlife surveys in the world." They also liked to note that the center's assessment team had accurately predicted the recent collapse of the New England groundfish populations.

Some NMFS critics, however—mostly fishermen but lately a few scientists as well—argued that the agency's assessment method, relying so heavily on a thirty-five-year-old collection regime and industry-collected data that everyone knew were suspect, left out so much information that it might be overstating the collapse or, equally serious, misreading its dynamics. It also bugged a lot of people, especially fishermen, that NMFS's math-heavy assessment models left fishermen virtually no way to contribute the knowledge they possessed—relatively unorganized, mostly qualitative information that NMFS tended to reject as "anecdotal." All of which (the argument went) might be crippling the effort to revive the fishery and to create more balanced, less punishing fishing regulations.

These objections were not merely academic. They were inspired in urgent fashion by the increasingly harsh and complicated regulations that the center's assessments had led to over the past few years. The Science Center didn't actually write the regulations, as most fishermen knew. That was the job of the New England Fishery Management Council (NEFMC), a twenty-one-member council (seventeen of whom voted) composed of industry, regulatory, and other representatives appointed by the secretary of commerce.[*] The center's job was merely to provide the council and its various technical committees with stock assessments, recommended sustainable fishing levels, and, on request, suggestions for reaching those goals.

---

[*] A branch of the National Oceanographic and Atmospheric Administration (NOAA), the National Marine Fisheries Service is part of the U.S. Department of Commerce.

This regulatory scheme, which had been optimistically termed "cooperative management" (meaning the scientists, industry, and other stakeholders would cooperate in managing the fishery), had been established by the 1976 Magnuson Act, a well-intended law that generated some unfortunate results. Passed to end overfishing of U.S. coastal waters by huge, foreign factory trawlers, Magnuson had two goals: to "Americanize" the U.S. coastal fisheries by extending the U.S. exclusive fishing zone from 12 to 200 miles offshore and by creating tax and investment incentives to expand the country's aging fleet; and to ensure a sustainable fishery by establishing eight regional councils that were supposed to regulate the fishery. The 200-mile limit chased off the foreign boats (the easy part), and the investment incentives, if anything, proved too effective, expanding the New England fleet, for instance, to twice its sustainable size in less than ten years. But the experiment in cooperative management foundered, particularly in New England.

Most explanations of the council system's failure emphasize a fox-guarding-the-henhouse dynamic, pointing out that the substantial industry presence on the regional fishery management councils (with most of the members appointed by the secretary of commerce from industry-heavy candidate lists offered by each region's governors) practically guaranteed inadequate restraint. How could we have expected fishermen to restrain fishing?

Such self-interested resistance to regulation, encouraged energetically by the most aggressive and short-sighted segments of the industry, definitely has played and still plays a role in the failure of most councils to restrain fishing. However, the system failed also for the simple reason that these clumsy, rather slow-footed regulatory bodies found themselves overwhelmed, underarmed, and frequently confused by the rapidly deteriorating situations they faced in the 1980s and 1990s.

Every council had several members drawn from the state marine fishery agencies in the region. If it was not obvious before, it seems well-established now that a regulatory body composed of roughly equal numbers of regulators and industry representatives will have trouble acting boldly or nimbly. Yet almost from their beginnings the councils faced a fast-blooming crisis that demanded quick action. Along with rapidly accelerating rates of overfishing (which were partly due to the incentives created and strengthened by Magnuson and the debt-driven expansion of the 1980s), their challenges included the complicated, decentralized nature of both fishery ecology and the fishing industry; the extreme difficulty of predicting the effects of any given management measure; a toolbag, put together by precedent and Congressional action, that offered mostly unwieldy or ineffective regulatory measures; and the doubt, in the face of large catches and seemingly abundant fish, that many fishermen held about NMFS's warnings of population decline. It didn't help that the councils met only about a dozen times a year.*

As a result, the New England council (and others) responded clumsily, tentatively, and slowly. Lacking both tools and conviction, it failed to control fishing adequately when the Science Center began warning of trouble in the mid-1980s, and it dickered and tinkered even when catch rates began plunging in the late 1980s. In fact, the New England Fishery Management Council (NEFMC) didn't act decisively until after the Boston-based Conservation Law Foundation sued the Secretary of Commerce and several NMFS officials for abdicating their legal responsibilities to conserve the fishery. By the time the consent decree from that suit compelled action, in 1994, so few groundfish remained on Georges Bank, which had seen the worst of the overfishing, that the only way the NEFMC could establish the newly required sustainable fishing levels was to close much of Georges Bank (an unprecedented and symbolically powerful step) and radically reduce the number of days fishermen were allowed to hunt groundfish elsewhere over Georges and in the Gulf. Thus began an ever-tightening series of constraints on fishermen's movements, time at sea, fishing options, and income.

The ever-increasing regulations inflicted a withering economic and emotional toll on most of New England's fishing families. While some fishermen recognized this as payback time for years of excess, many others,

---

*That the councils' recommended measures could be accepted or rejected by the secretary of commerce, but not modified, encouraged approval of weak measures. Recognizing this, the revised Magnuson Act of 1996, known also as the Sustainable Fisheries Act (SFA), gave the Commerce Department the authority to impose its own management plans and regulations if it found a council was failing in its assigned task of protecting the fishery.

having advocated for themselves too well when solutions might have come more easily, reacted with fury at the council for its harsh restrictions and at NMFS's scientists for bearing bad news. In the blurring of boundaries and denial of limits that the spectacularly dysfunctional council process encouraged, this louder, angrier sector of the fishing community didn't want to bother distinguishing between NMFS's science and policy branches, and some found pleasure in the idea of killing the messenger. Now NMFS scientists were bringing yet more bad news: The latest assessments had found not only that Georges' groundfish were recovering with excruciating slowness but that Gulf of Maine groundfish, which were getting hammered by all the boats that had been kicked off Georges, were also in danger of crashing, and could be saved only by sharp new fishing restrictions.

All largely because of what the *Delaware* and the *Albatross* were catching and counting on the assessment cruises. An awareness of these implications quietly permeated the cruises. In a dismal way this was an encouraging change from five or ten years previous. Through the late 1980s and early 1990s, the *Delaware* and the *Albatross* ran their surveys amid fleets of commercial boats fishing without restraint—fishing, it must have seemed, in defiance or even disdain of the Northeast Fisheries Science Center's findings. Now, at least on Georges, the NMFS vessels set their nets in waters emptied by the regulations their work had spawned. If the NMFS staffers on these cruises felt besieged by the industry's criticisms, they could at least work knowing that everyone would pay attention (even if not happily) to what they found.

Still, the notion that by miscounting the fish we could needlessly make fishermen yet more miserable was disturbing to contemplate. Which was why I was out there. I'd heard NMFS explain why its work was accurate; I'd heard fishermen explain why it was garbage. I wanted to see for myself how the Science Center counted fish and how it made what it made of what it found—what, in short, it had made of Bigelow's work. I also wanted to see firsthand what NMFS would find in that critical year of 1998, when the Gulf of Maine cod were supposedly crashing, and the fishermen were going broke and talking murder, and everyone who hadn't managed to find insulation from blood, sweat, tears, and fish scales was frustrated and angry.

By the late 1990s, of course, every year was critical. A decade of dismal assessments and ruinous restrictions had left everyone—fish, fishermen, regulators, scientists—with little wiggle room.

It made one rather careful about identifying the fish.

"Oh, you definitely feel the pressure out here," Galbraith told me later that first day, when we had sorted the first catch and were passing the time in the *Delaware*'s wet lab before the next haul. "Some of these fishermen say we don't care, or that we're sloppy. But we take a lot of trouble to get these things right. Like this calibration cruise. The whole purpose of having the *Delaware* out here right now is to make sure we're counting these fish accurately, so each year's data are as reliable as possible." He was referring to the fact that the *Delaware* was running alongside its sister ship, the *Albatross,* for the sole purpose of running duplicate tows. The *Albatross* was taking the actual survey, collecting the numbers that would be used (along with some catch data from industry) to create the next assessment. We on the *Delaware,* meanwhile, would run a calibration cruise, sailing an identical course and making identical tows so that the Science Center could see whether some recent changes to the *Delaware*'s fishing gear (new net doors and winches) had altered the rate at which the *Delaware* caught fish. Two dozen crew members and ten grand a day just to calibrate the giant scale that was the *Delaware*. To pull the data tighter yet, we were testing a set of ultrasonic monitors that, secured around the mouth of the net, would send to the ship an image of how the net was riding in the water. The monitors were sitting on the steel counter next to Galbraith, four red plastic cylinders a bit larger than liter soda bottles. Dan Doolittle, a bright young former turtle researcher who was volunteering on his second consecutive survey cruise, had taken on the project of testing the new devices; he hoped to send them out on the next haul. Also on the counter were a stopwatch and clipboard the watch chiefs used to time how long it took the net to be let out and retrieved at either end of each half-hour haul—information they would compare to similar data from the *Albatross* and the previous winch rig on the *Delaware*.

"You get hammered enough, you get defensive about these things," said Galbraith. "One time a few years ago we had this huge controversy about the clam survey because we had to toss out this one cruise's unusually high clam numbers after we realized the boat had been towing too fast and scooping up too many clams. So we had these numbers making it look as if there were lots of clams, and we essentially threw them out. The fishermen were furious. Said we were fudging the numbers because we didn't want them to fish. It was a nightmare. A huge stink. So on the next cruise, there was unbe*lie*vable pressure to demonstrate that the tows were done exactly right. I mean, we calibrated *everything*. It was intense—twelve days, everybody tense, hell breaking loose back on shore, the scientists on board bickering with the fishing crew about every little thing about the dredge set-up, bickering with the boat crew about the speed and depth of every set, looking over their shoulders. It was awful. That was a *long* cruise.

"But when you get criticized all the time—now we get *sued* all the time—you start questioning everything. Which is probably good. I mean, we should be careful. These fishermen are getting killed. I've even wondered a couple of times whether maybe we're catching fewer fish because the fish are starting to recognize the boat's acoustic signature—the distinct sound the boat makes going through the water—so we really *are* undercounting fish. Which makes no sense at all—I mean, going out twice a year, we don't catch enough fish to select for fish that would avoid the boat's signature. There's almost no way that could be. But you start wondering about these things.

"The other thing you hear from fishermen is we undercount fish because we use ancient gear. Which is true. I mean, it's true we use old gear. We use gear from 1963, called a number thirty-six Yankee otter trawl, because that's what we used when the *Albatross* started these cruises, so we have to keep using it to get consistent results. No one uses those nets anymore. You can't even get them. We have to have them specially made. But we do it because if we want a reliable census, we have to count the fish the same way every year. We're always doing all this stuff to make sure everything's the same. Like when all the netmakers switched from manila to polypropylene for the net material. We did all these calibration studies using two nets, a manila net and a polypropylene net, to create adjustment factors for the new material, see if it fished any different. They worried that the polypropylene nets might push a different sort of compression wave in front of the net and affect how many fish we caught. So we checked to see if we needed to adjust for that, too. We check *everything*. Don't we, Jorge?"

The shift's two fishermen, Gene Magan and Jorge Barbosa (who pronounced his name "George" rather than in the Hispanic manner), had come into the room while we were talking. Both the *Albatross* and *Delaware* had professional fishing crews that handled the specialized, rather dangerous work of setting and retrieving the nets; like most of those fishing crew members, Barbosa and Magan had both fished commercially for years before hiring on with NOAA. Despite some standard disgruntlements, Barbosa and Magan seemed to enjoy their jobs well enough. Barbosa was a fit-looking forty or forty-five, Magan a few years older and with a physique and bulgy, expressive eyes reminiscent of Rodney Dangerfield. I kept expecting him to say he got no respect.

"Absolutely," said Barbosa, smiling. "Makes a big pain in the ass."

He nodded at my notebook.

"You're the guy writing the book."

I said I was.

"What's it about?"

I told him that at this point it seemed to be about how fishermen and NMFS disagreed on whether there were any fish out here.

"Oh, there's fish out here," said Barbosa. "Dogfish." He and Magan and John all laughed. The dogfish, an odious, universally hated small shark, had become much more common than the cod and haddock everyone would rather catch. Even though they had become by default a vital part of most fishermen's income, fishermen hated them as much as the NMFS crews did.

"But other than dogfish," said Barbosa, "no. Not as many fish as should be."

"Things have definitely changed," said Magan. "It's crazy to even argue about. There aren't nearly as many fish as there used to be. We used to bring five, eight, even ten thousand pounds a haul, all the time. That was standard. Big haul was ten or fifteen thousand pounds. You went out for five or six days, you'd bring in sixty to a hundred thousand pounds of groundfish. Now a fifteen-thousand-pound trip is thought of as good. I think the younger guys, who never saw those catches ten or fifteen years ago, they think things are okay now. But there's nowhere near as many fish out there now."

"I had one trip," said Barbosa, "I brought a hundred thirty thousand pounds into Gloucester. Wasn't every trip like that. But used to be good trips. Not like this where you bring in just a few here and there. These guys fishing now, they think it's great if they bring in twenty thousand pounds after a week of fishing."

"Only way they make money is they work small crews," said Magan. "These big boats that used to have twelve, thirteen guys, they take five or six now. Ones that used to take six or seven now take two or three. They don't need any more than that because there aren't any fish."

"What do your friends in New Bedford say when you tell them all this?" I asked.

"They don't want to hear it," said Barbosa. "Some are going broke; some are doing okay. They all wish they could fish more. They're allowed eighty-eight days fishing a year; it's not enough. They feel if they got more days and caught the same number of fish every day they catch now, they'd be doing well. They think the fish are okay and it's the regulations that are killing them. But the fish aren't okay. There are just not too many fish out here."

"So what should be done about all this?" I asked.

"Shut the whole thing down," said Barbosa. "Charge a tariff on imported fish, use the money to subsidize the fishermen. Train some for other jobs; pay the rest to do research."

It was about as good a solution as I'd yet heard.

3

If you want NMFS's explanation of how this whole mess occurred, the best place to go is probably Steve Murawski's office at the Science Center in Woods Hole. The center resides in a group of buildings only feet from Woods Hole harbor, on an outside corner of land with water surrounding the ell the two buildings form. Perhaps appropriately, Murawski's office and those of the rest of the assessment team are in a different building from the offices of Galbraith and the other fish counters in the Resource Survey Branch, separated from them by a parking lot, two flights of stairs, and the center's library—elevated above the resource survey folks, as it were, and on the far side of formalized knowledge. As head of the center's Population Dynamics Branch, Murawski is dedicated not so much to catching or even counting fish as he is to taking the sums arriving from several sources—primarily the "indices of relative abundance" produced by the Resource Survey Branch survey cruises and catch data from industry (such as dealer reports and standardized logbooks that the fishermen keep, with an accuracy everyone acknowledges is open to question), in some cases supplemented by data from state fishery agencies or other sources—and pushing them through some intimidating equations and computer models to produce the center's stock assessments.

Given this job, Murawski's office is exactly as it should be. Whereas Galbraith's tiny cubicle looks out on Woods Hole harbor (where the sight of bonito ripping the surface sometimes tests his dedication), Murawski's window faces landward, showing a street, a tennis court, and the town's two great private research institutions, the Marine Biological Laboratory and the Woods Hole Oceanographic Institution. In a building half-surrounded by water, Murawski sits inside almost completely surrounded by data. His office walls are lined with cabinets and shelving stuffed with books, data printouts, analyses, scientific papers. More reports and printouts cover his three desks. His computer, holding several gigabytes more, glows on the desk beneath the window. The day I visited him, we sat at a library table in the room's center. It was a table at which eight could have sat comfortably for dinner were it not covered with more books, reports, and stacks of data. Murawski sat back in his swivel chair and tried to explain to me how all this information failed to save the New England fishery.

"You can imagine we've gone back and traced the history of our numbers and our advice to the New England Fisheries Management Council," he told me. "It was about '85 we started warning, pointing out that exploitation rates had grown to about twice what they should be, then even more. Every year our warnings got more intense, more pointed. No one paid any attention until the bottom fell out."

By that time, the early 1990s, even many fishermen were acknowledging that Georges' groundfish populations were crashing, for after years of fishing harder and catching more, they were fishing harder and catching less. In Murawski's terms, their catch-per-unit-effort, or CPUE—the total weight of the fish they caught in an average day—was dropping. Murawski's trump card in the argument over whether the center's ground-fish numbers were accurate is a graph with two plotted lines, one representing the CPUE, the other representing the survey-cruise catches. The two lines—one based on the fishermen's catches, one on NMFS's—follow each other's contours perfectly through the period. Both lines rise in bull-market fashion into the mid-1980s, dip modestly, rise again toward the end of that decade (the result of a particularly successful groundfish spawning in 1987), and then plunge steeply toward a nadir in 1992, from which both lines more or less flatten through the 1990s.

The cause of this fall, said Murawski, was quite simple—a variation on the commonly cited "Too many boats chasing too few fish."

"The fleet was roughly twice the size the fishery could stand, and most of the boats were fishing hard. We were catching two, three times the number of fish the populations could stand. That's saying a lot. A healthy population of groundfish can sustain a surprising amount of fishing, because healthy groundfish populations produce a *lot* of eggs. A healthy adult female cod, for instance, typically puts out one to two million eggs at each spawning. That's an average. Older fish put out a lot more, maybe five to ten million. But average, about two million. But few of those eggs make it. Even in a banner year, where everything goes right for the fish, only a slim fraction of the eggs—maybe a hundredth of one percent, or .01—survive the first year. The rest fall to all these disasters that are waiting to happen to them. They don't hatch. They get eaten. A current sweeps them off the shallows and into the deeps, so they hatch into a place that has no food. These things make first-year survival rates vary, on a yearly basis, by a factor of ten or more. That's a lot. It means the difference between a female's million eggs producing a hundred fish that live to year one and that million eggs producing only ten fish that make it that far. And those year ones still stand vulnerable to many predators—including bigger cod—because they don't get really big till about their third or fourth year.

"Still, they do all right if there's enough of them and they have a little luck. If they produce a good year-class every few years—that is, if every two, three, or four years they have a year where a lot of the fish survive to adulthood—they do okay. They can withstand a surprising degree of fishing pressure. In fact, you can catch between a quarter and a third of adult cod every year and still have a sustainable population. That's an incredibly high rate.

"Problem was, during the eighties we were catching between sixty and eighty percent of the main groundfish species here—the cod, haddock, and yellowtail flounder. Boggles my mind that we could hunt something that successfully—that we could catch six or eight of every ten adult groundfish in the whole ecosystem. That's way beyond the fishes' capacity to regenerate. I mean, even at sixty percent there's no way they can make it. For every ten adult animals, fishing boats are taking six? Two more will die of natural causes. That leaves just two—if you're lucky, one male and one female—to replace the stock.

"That's not going to work. I mean, take cod. You've got a fish that can live twenty years, and on average, without fishing pressure, lives about nine or ten. A female that lives that long will spawn four or five times. The first time or two, when she's five or six, she's not very good at it. She lays only a few eggs, and they're not very big or robust, so very few of them even hatch—maybe thirteen percent the first year and sixty the second. So through her ten-year normal average life she'll have maybe four or five effective spawnings.

"But we were fishing these fish so hard, catching six to eight out of ten adults, that we were catching most of those female fish almost as soon as they reached adult size—before they got good at spawning. We effectively decreased their average age to about five or six. That meant they were averaging *less* than one effective spawning per lifetime. And this is a species that counts on having four or five effective spawnings per lifetime.

"Think about it. You've got an animal that normally spawns multiple times so that if she doesn't get her genes in the pool this year, she will the next year or the next or the next. Because that's the only way she can account for the huge variabilities of the marine environment—for all those disasters that can happen to her eggs and young. We took that away. We took away the entire evolutionary strategy of our most valuable species."

Had the industry and the council reacted sooner and more firmly to the center's warnings, says Murawski (and many others), the crash on Georges could have been averted. Dave Crestin, a retired NMFS staffer, recalls a meeting in the late 1980s when his friend Vaughn Anthony, a leader on the assessment team at the time, was hooted and laughed at when he tried to warn a roomful of fishermen of the coming dive. "They just didn't want to hear it," said Crestin. "He's one of the best assessment scientists we ever had, and they laughed in his face."

Murawski sees the industry's resistance to NMFS's warnings in more scientific, arguably more charitable terms.

"I think there was a belief by the council and many others then that fish and fishermen are in a long-term predator-prey cycle. You remember the classic example of the lynx and the hare—that they go through long-term sine waves that are out of phase, the hare populations going up, the lynx responding by growing in number, then eating too many hares so the hare population crashes, and then lynx go down again, and then the hare rebound and it starts over. A lot of people thought that fish and fisherman were in that sort of cycle—that when the fish went down, some fishermen would go out of business, relieving the pressure, and you'd get this sort of self-regulating fluctuation around some mean level of harvest."

As Murawski pointed out, that model did not take into account that bankrupt fishermen would sell their boats to other, not-yet-broke and more optimistic fishermen, who (encouraged by the special tax incentives the industry enjoyed) would put yet more sophisticated fish-finding gear on the boats and go fishing with renewed vigor and better technology. The lynx population, in other words, had reserve units of rested, healthy but hungry lynx, and thanks to ever-improving hare-finding technology and gear, ever-more powerful eyes and claws.

In retrospect, says Murawski, "It was a surprisingly naive model the managers were operating under. I think they were sold a bill of goods, frankly, by people who don't like to see the hand of government in private business. That's a strong feeling. People go fishing, it's a hard life. They do it because they like to be outdoors. And they have this idea that once you shove off from shore, you're master of your own fate. It's powerful imagery. Unfortunately, it falls apart when you fish with the sort of technology they have these days.

"This thing could have been averted. All our simulation shows that if a rational fishing plan had been put in place, we'd have a viable stock right now. These guys would still be out catching lots of fish."

Murawski is a whip-smart, articulate, persuasive explainer, but you don't have to take his word on all this. A few months after I talked with him, just before I sailed on the *Delaware* for Georges Bank, a special National Research Council panel published a report, *Review of Northeast Fishery Stock Assessments*, that essentially confirmed his account.[2] The panel, composed of highly respected fishery scientists not connected with NMFS, had been convened at the request of Congress in response to industry's doubts about the accuracy of the center's work. It found that the Northeast Fisheries Science Center produced valid assessments that justified the strong restrictions that had disrupted the industry. It also specifically exonerated the NEFSC of some weaknesses in study design and method that a separate NRC report, published only weeks before, had found in other NMFS regions and assessment programs.[3] The panel further agreed that Georges Bank cod, Gulf of Maine cod, and yellowtail flounder—stocks that the center had identified as being in danger of collapse—would likely crash if fishing restrictions didn't tighten even further.

Murawski had gotten wind of these findings by the time I talked to him and, understandably, exuded an almost palpable sense of relief regarding the report. Like many of the center's staff, he was quite conscious of working in a town that had a heritage as a leader in oceanography and fishery science; it must have sickened him to see the charges that the earlier, more generally and nationally focused report had enumerated. "Absence of adequate data," "stock assessments ... compromised by incomplete or variable data," "the biomass of recovering populations tends to be underestimated": These are ugly things for fishery scientists to hear said about their work, and good things to be absolved of. The new report meant that Murawski and his colleagues would not have to drive past the East's greatest private centers of marine science, the Woods Hole Oceanographic Institution, founded by Bigelow, and its close cousin the Marine Biological Laboratory, to work in a government agency that had been convicted by its peers of shoddy work.

Yet anyone at the center who had been expecting the industry to say, "Oh, sorry. You were right all along," would have been disappointed. In fact, in the same month that the NRC's *Review of Northeast Fishery Stock Assessments* came out, the industry's leading trade journal, *National Fisherman* (which seemed to relish whacking NMFS whenever possible), ran a cover story on the first NRC report—the one levying fairly serious general charges at some of NMFS's assessments—without mentioning that a separate NRC report specifically on the Northeast assessments was forthcoming. The Northeast report would have been available to *National Fisherman*'s writers and editors before press time, and they either knew about it or should have known about it. Omitting it from their cover story, however, allowed them to apply the criticisms from the earlier national report, including a statement that NMFS based some of its assessments on survey data taken only every three years, to the Northeast assessments—which, because its assessments had found the highest levels of overfishing and thus produced the harshest restrictions, were a primary focus of the industry's growing campaign to disparage NMFS. As a result, New England fishermen angry with NMFS would often repeat the March 1998 issue's assertions of sloppy science and cite the first NRC report (ignoring the second) as proof that the center's work was junk and that the scientists "didn't know the back end of the boat," as a common complaint among fishermen had it. By this time, of course, no one at the center expected charitable treatment from *National Fisherman*. They had to take what comfort they could in being confirmed by their peers.

Any other finding from the NRC, actually, would have been a long fall for the center, for the assessments coming out of NMFS's Woods Hole office had for some time been considered among the world's best. The Northeast Fisheries Science Center's work was a dead-ahead, state-of-the-art application of mainstream fishery science that drew quite self-consciously on the institutional, human, and data foundations established by such giants of American fisheries science as Henry Bigelow and Spencer Baird, who founded the Fisheries Bureau (later NMFS) in 1871. Having grown directly from the long-term intensive-area study approach used by Bigelow in the Gulf of Maine and the International Council for Exploration of the Sea in Europe, the center's methodology incorporated the best of those traditions and some impressively sophisticated computerized modeling that let researchers double-check their predictions and "backcast" to make sure current findings jibed with previous assessments, creating a sort of self-correcting loop. In short, as the NRC report asserted (and I witnessed), the scientists of the Northeast Fisheries Science Center collected data with rigor and consistency and analyzed them with powerful and well-established statistical methods. They were not, as some fishermen accused, "just makin' this shit up."

4

We took our time getting to Georges, first looping south to do a set of stations below the Great South Channel, then moving up through the Channel to test the waters around Cape Cod and Cape Ann. The *Delaware* followed the *Albatross* as we played the oceanic game of connect-the-dots required to hit the randomly spaced sampling points that had been generated by the NMFS computer back on shore. Our jagged course was set by Galbraith's colleague from the Resource Survey Branch, Linda Despres, who, sailing as chief scientist of the *Albatross,* worked out with *Albatross* captain Jack Moakley how best to connect the dots.

It was pleasant labor, sliding along over long, soft swells and sorting fish every hour or so under high skies. With the *Albatross* crew doing all the dissection work, we on the *Delaware* had only to count, measure, and weigh our rather small catches at a leisurely pace. We had time to read, talk, scan the skies for birds, nurse a cup of coffee. It was hard not to drink a lot of coffee, for the strange shifts quickly produced a glaze of fatigue. I was lucky enough to draw the day watch, which meant I worked from six to noon and six to midnight, and thus could get about five hours' sleep at a more or less normal time, from just after midnight to 5:15, and I could nap in the afternoon as well, when the twelve-to-six watch was on. Some people on the night watch, however, developed impressive heads of fatigue. One evening I found Dan Doolittle sitting before one of the computer terminals studying a spiral notebook and intermittently tapping at the keyboard, apparently entering data from the notebook into a spreadsheet. When I idly asked what he was up to—meaning what sort of data he was working on—he looked at me foggily, inhaled and started to say something, couldn't find any words, took another breath, let it out, and then said, "I'm, uh, I'm putting these, um, these numbers—" he patted the note-book, "into this, um, uh . . . ." He sighed, patting the top of the computer monitor.

"Computer," I suggested.

"Yes," he said, smiling and nodding his head. "Thanks."

On warm-weather cruises, Galbraith explained, people brought lawn chairs and beach towels or even hammocks so they could catch some rays on slow days like these, and on tandem cruises they sometimes rigged giant water-balloon slingshots with which to mortar the sister ship.

No such sunny games in early April on Georges Bank, however. Indeed, the Bank asserted its reputation for lousy spring weather as we neared it on the third day. The bright skies we'd sailed under while work-ing off the Cape dimmed as we headed south, lowering and darkening until we were sailing through a horizonless fog. Visibility ranged from a half-mile to a mile, keeping the bridge crew on their toes, eyes to the radar. The winds rose into the twenties and thirties, pushing the broad 3-foot pil-lows of the first few days into steep, frothy, 8-to-10–foot hillocks. The boat dived, rolled, yawed, and sometimes banged its way along. To some of the volunteers the rising seas brought seasickness' singular collection of symp-toms—a dark, sapping queasiness under the sternum; a vague and then growing desire to lie down; and finally a stalking, demoralizing nausea. The most stricken picked through their meals silently, then grayly climbed the stairs to count fish or look for a place to lay their heads. You weren't allowed to return to your stateroom during your shift, lest you disturb the sleep of your off-shift roommate, so volunteers had to seek rest on the chairs and steel tables in the lab or the booths in the mess, the only real public areas on the *Delaware*. Between meals you could find the afflicted lying in the mess' booths, cradling their heads in their arms.

Work, generally the best antidote to seasickness, did not offer as much distraction as it might have, for we caught pathetically little. With tows only a half-hour long, you didn't expect huge catches. But we were bring-ing in loads you could fit in a laundry basket. Few of the fish we caught were gadids, even though at that time of year Georges should have been teeming with spawning cod and haddock. The gadids we brought up were older, suggesting a lack of the juvenile fish that might constitute recovery. "Not good," became the refrain when the fishing crew hauled up the net.

We had to resist, of course, the temptation to draw any firm conclusions from the tiny, dripping loads that bounced into the checker. We all recognized, as scientists trained and in training, that no individual haul meant anything significant in itself, and that even a full day of mingy hauls didn't mean that much; only the overall pattern meant anything, and even it had meaning only because of how it fit into the larger, longer population curves. Yet it was impossible not to feel worried as the sparse hauls added up. After all, each haul did mean something, the same way every pitch in a baseball game means something. It was a long game, but the cod team was putting very little on the board.

The third night out we crossed over the rim of the Gulf onto Georges and in heavy seas began working a widely spaced array of stations along the Bank's northern side. I woke, showered, breakfasted, and went to see if the net was out. I found Barbosa just outside the door to the wet lab.

"Goes out in a few minutes," he said.

It was a close, cold, gray day. Barbosa and I stood in the sheltered work space just outside the lab doors and looked at the foam-flecked, pewter-gray water off the stern. The boat climbed and slid through the lumpy hillocks, which blurred in the distance to meld into the sky's barely brighter grayness. There was no horizon. I looked around and could not tell where the sunrise had been. Above us the rubber gloves and woolen glove liners of the scientific crews, hung on a line to dry, swung erratically back and forth.

"They catch much last night?" I asked Barbosa.

"Not much. Some haddock."

"Where are we?"

"Cultivator Shoals. I had some good catches out here." Cultivator, a part of the Bank's high, shallow, north-central section, had for centuries been one of the Bank's richest parts.

"You mean when you were fishing?" I asked him.

"Yes," Barbosa said, smiling. "Not so much lately."

An hour later, as Galbraith and the rest of us looked on, Barbosa and Magan landed three tons of *Squalus acanthias*—the dreaded spiny dogfish. The net creaked as it swung ponderously over the deck, and when Barbosa had banged out the pin that held the cod-end's puckerstring and yanked the string loose and jumped aside, the fish gushed out into a gray heap roughly the size and shape of a Volkswagen beetle. Barbosa said, "Oh boy," and then, kindly (for it wasn't part of his job to sort fish), pitched in to count the fish. I quickly came to understand all the disparaging things I had heard about dogfish. The creatures were only 2 feet or so full grown, and their teeth weren't much to worry about, but they lacked charm and, after a few minutes, interest.

Even Bigelow found dogs objectionable.

> Much has been written of the habits of the spiny dogfish, but nothing to recommend it from the standpoint of the fishermen or of its fellow creatures in the sea. . . . Voracious almost beyond belief, the dogfish entirely deserves its bad reputation. Not only does it harry and drive off mackerel, herring, and even fish as large as cod and haddock, but it destroys vast numbers of them. Again and again fishermen have described packs of dogs dashing among schools of mackerel, and even attacking them within the seines, biting through the net, and releasing such of the catch as escapes them. . . . Often, too, they bite groundfish from the hooks of long lines, or take the baits and make it futile to fish with hook and line where they abound.[4]

Counting and sorting them made ugly work. Each dogfish had a sharp, 2-inch spike half-hidden just in front of each of its two dorsal fins, and in the net bag's scrambled compression, they had stabbed and bloodied each other with the spikes, making the whole pile a putrid-looking mass of gray, white, and red. Some were flattened from being squeezed into the overfull net. They had the classic shark frown and affectless shark eyes. They didn't smell too good, either. Looking at the pile, Galbraith declared that we would take a subsample, filling ten baskets with fish that we would sex and weigh and measure, then merely counting the rest as we threw them overboard. From the measured subsample we would extrapolate the sexes and sizes of the unmeasured.

We quickly filled ten baskets, then set about counting and pitching the rest. Flying dogfish filled the airspace above 3 feet. To avoid being spiked—dogged—I knelt at the edge of the pile with my back to the rail and started grabbing fish two at a time, one in each gloved hand, taking care not to grab those spikes, and then flinging them straight back over my shoulders to sail over the rail. In about four minutes I counted and flung 248 dogfish.

When we weighed the subsample and did our math, we found we had handled about three tons. (We later learned the *Albatross* dealt with a similar load.) We had to hose each other down to clear our foul-weather gear of all the blood.

We were not alone in our disgust. For fishermen, dogs had long constituted a sort of fool's gold, filling nets that would be brought to the surface excitedly only to be found full of these foul fish. The fishermen could and did sell them (if you've dined on "Cape shark," you've eaten dogfish) but only for a fraction of what they got for cod. They resented them, too, on aesthetic grounds: It disappointed mightily to look for the lovely olive and bronze tones of the soft-skinned cod and encounter instead the drab, rough gray of this flaccid-fleshed fish.

More recently dogs had become a sort of fool's lament, as they and their fellow elasmobranches, the skates, had displaced gadids and flounders as the Northwest Atlantic's dominant fish. For as far back as records go, gadids and flounders—the heart of the groundfishery—had outnumbered dogs and skates by factors of two to ten. But when we decimated groundfish in the mid-1980s, dogs and skates, apparently enjoying the lack of competition for food and space along the bottom, exploded into the niche left by the absent groundfish. The abundance curves of these two groups crossed, and by 1994, dogs and skates outnumbered groundfish by eight to ten times. The neighborhood went completely to hell.

The dog's own abundance had since declined somewhat as *it* became overfished, but it still outnumbered gadids, and the bloody, gray, spiny *Squalus* now symbolized the disruption of the New England marine ecosystem. This shift in population, and the troubling failure of New England groundfish, especially cod, to rebound even in the protected areas of Georges Bank, had led some to wonder whether overfishing had created some fundamental change in ecosystem structure that would prevent the groundfish's return—a "regime shift." A variation on the general regime-shift hypothesis was the "fishing down the food web" theory. The fishing-down theory held that overfishing was eroding the top of the food-web pyramid by repeatedly fishing the pyramid's top predators into or near commercial extinction (the point at which they're so rare they're no longer worth fishing for), then moving down to the next level of predators on the food pyramid and then fishing them out, and so on, in a repeating downward path that simplified and impoverished marine ecosystems.[*] A team of researchers led by a smart young number-cruncher named Daniel Pauly had given this theory its most elaborate and convincing airing in a paper published a few weeks before the spring 1998 Georges cruise.[5] Examining forty-five years of worldwide fishery landings records, Pauly and his colleagues found that the industry had measurably worked its way down the food web over that time, fishing to relative scarcity many bigger, top predators such as tuna and cod and red snapper and then aiming its nets and hooks further down the food web at species such as shrimp, herring, or pout.

Pauly et alia's main evidence was a drop from 1950 to 1994 in the total catch's average "trophic level," as measured on a 1-to-7 scale that assigns predators at the top of the food web values in the upper range and strictly vegetarian invertebrates such as zooplankton a 1. (Adult bluefin tuna and cod have trophic levels around 5, dogfish around 4, Atlantic mackerel around 3, and Pandalid shrimp, the most common in New England waters, around 2.)[6] They found that whenever a region's industry turned its attention from an overexploited high-level species to an abundant fish lower in the food web, the industry's total catch (in weight) would rise for a while, then drop again as overfishing depleted the new target species—after which boats would then target yet another new species even lower on the chain. This dynamic appeared most pronounced in the most developed fisheries. The Northwest Atlantic, for instance, had seen a drop in trophic level of 10 percent as the industry had fished the region's tuna, swordfish, and now cod and haddock into relative scarcity and then concentrated more on lower-trophic-level species such as shrimp, squid, and herring. (The turn to dogfish had if anything moderated this trophic-level decline, since dogfish have a relatively high trophic level.)

Beyond this top-down erosion, Pauly noted, the move down the food web could disrupt the ocean ecosystem in more subtle ways. In the North Sea, for instance, fishermen banned from fishing that country's decimated cod stocks began targeting ocean pout, cutting their numbers enough to

---

[*]The terms *food web* and *food pyramid* are now preferred over *food chain* for their recognition that most species eat and/or are eaten by more than one species.

allow an *increase* in the population of one of the pout's food staples, small krill, which then started competing more effectively with cod larvae for plankton. The turn from cod to pout therefore indirectly stressed the larval cod on which the cod's recovery depended.

Left unchecked, Pauly and his colleagues argued, this downward spiral would likely extinguish ecosystems and entire fishing industries. As their paper put it, "There is a lower size limit for what can be caught and marketed, and zooplankton is not going to be reaching our dinner plates in the foreseeable future."[7]

Some scientists objected to Pauly's analysis and conclusions, arguing that the differences in what fishermen were catching between 1950 and 1994 could be the result of changes in demand as well as in actual fish populations. But you didn't have to swallow whole the fishing-down-the-food-web theory to feel alarmed about the dynamics it described. Anyone familiar with the New England fishery could see that something similar might be at work in these waters. Whether it was the dogfish's usurpation of the gadid's dominance or the cod's continued failure to have a strong reproductive year (cod usually have three or four strong year-classes a decade, but New England populations hadn't had one since 1987), something was seriously amiss.

At lunch, after we had changed out of our foul-weather gear and washed the dogfish off our hands, I asked Galbraith what he thought about the regime shift and fishing-down-the-web theories.

"Well, I don't know about all that. I'm not really up on it. But things have certainly changed, and the lack of rebound is really disturbing. Cod's the most obvious. And it affects everyone who lives along the coast. Fishermen talk as if it's just their own interests they damage when they overfish, but it affects everybody. Everybody who cares about fish, anyway. I mean, there's a whole mess of fish that you used to be able to fish for inshore that just aren't there anymore. Used to be you could fish inshore for cod and have a good afternoon of it. Now it's hopeless. Or scup. Everybody used to fish for those inshore. They're not there anymore."

I asked him what a scup was. He put down his fork and smiled. Galbraith almost always smiles when he talks about a particular fish—any fish except dogfish, anyway. "A scup is an interesting fish," he said. "You know what a porgy is? Same thing. Kind of narrow side to side and tall from bottom edge to top, perchlike that way, and it's got that spiky dorsal fin like perch do, so you have to brush it down from the front to handle it. Except it's silvery rather than any sort of yellow. Most of them are about a foot long." He laughed. "Not that much of a fish, really. It's a panfish, basically. But it's fun to fish for."

"Easy to catch," I suggested.

"Yes! Fight like crazy. It's what every kid starts out fishing for around here. Or used to, before we whacked the crap out of everything."

"So," I said, "no cod or scup. What'll you fish for this summer?"

"Actually—and I don't have a real scientific basis for this," he said, chuckling, "I really believe this could be a great year for bonito."

"Why's that?"

"Well, water's been warm. That sometimes does it. There seemed to be a lot last year. I've heard they're good farther south. And," he said, beginning to laugh, "because I really want it to be."

I considered telling him he sounded like a cod fisherman explaining why he thought there were a lot of fish out there, then realized there was no need. That was precisely what he was laughing about.

5

That afternoon we cut west toward the Great South Channel again, worked some stations on the Bank's northwest edge, and then steamed overnight across the Bank toward the southern edge. In my bunk, I hovered and bobbed through a night of sleep at once shallow and deep, feeling as if I were expanding and floating whenever the boat dropped, pulling the bunk down away from me, then denser and heavier as it rose again—a strange rocking motion that, perhaps because it wasn't compressing my organs atop one another, I found very comforting. Though I seemed to be pleasantly half-aware of this strange sensation the whole time I slept, half-aware, too, of dreams of cod and tiny, translucent squid, I woke feeling fresher than I had since leaving the gray hills of Vermont for Woods Hole.

When I woke it took me a few minutes to realize the seas had calmed. I jumped down from my high bunk (the first time in three days I hadn't had to climb down clinging to the bed frame to avoid being tossed by the rocking boat), showered, ate an omelet, and headed up to the lab. Out the window the skies had lifted, though they hadn't exactly cleared, and the sea had relaxed into rounder rollers instead of the stiffer, steeper waves we'd dealt with the day before.

Barbosa and Galbraith were standing just outside the doorway. "I wouldn't fish this," Barbosa was saying. Meaning he wouldn't risk the net on this craggy bottom if he were running the boat. We were over the middle of Georges Bank now, south of Cultivator, where millennia of strong rips had combed the shallows to leave long, knobby ridges that posed distinct hazards to fishing nets. Barbosa had owned and run his own 65-foot trawler for eight years before selling it (in a 1995 government buyback program aimed at retiring fishing permits), so he wasn't talking idly. You had to fish ground like that carefully, and perhaps not at all if you were using the small roller-gear that the *Delaware* had on the bottom cords of its nets. (The bottom cords of many newer nets had bigger, bouncier rollers that could pass over rough bottom more safely.) Of course, as Barbosa recognized, NMFS considered a different cost–benefit equation than he did when it decided whether to fish rough ground. A commercial boat risking several thousand dollars' worth of gear on toothy terrain gained only the possibility of a good haul—and rarely did a load fetch what a net cost. The *Delaware* and the *Albatross,* however, got their catch—a read on how many fish were there—regardless of whether the net came up stuffed or empty.

Unless, of course, the net came up with a huge hole or not at all, which sometimes happened. The net could hang and rip or tear loose altogether on a big rock or a spire or one of the wrecks that littered these shoals, boats that had run aground and broken up in storms. (These wrecks are marked on charts with sobering density.) When the net hung hard, you could tell, even inside the boat. The ship would lurch, the winches groan, and if the snag did not immediately release, the cables would start shuddering. The boat might swivel off course as it played tug-of-war with the sea bottom. Now and then the cable would part, and if it snapped near the surface the end could come bullwhipping up onto the fish deck. Things could actually get worse if the cables *didn't* snap, for if the net hung fast and the boat kept pulling, the cables could pull the stern down, and on a ramp-sterned boat like the *Delaware,* that could mean shipping water. That had happened once a few years previous. The *Delaware* and *Albatross* were running a tandem cruise on Georges when the weather got heavy, with a strong current surging over the Bank. To protect its stern ramp, the *Delaware* usually towed into such currents. The *Albatross,* however, lacked the power to tow into such a strong current, so both boats towed with it, rolling sickeningly forward as the following seas lifted and then passed under the boats. Then the *Delaware* hung fast on some bad ground. Before anyone fully realized what was happening, the straining cables began pulling the stern down. A swell broke over the ramp and washed over the fish deck, and as the wave pushed the stern yet lower the sea blasted through the open rear doors to the wet lab and entered the boat. Thousands of gallons of cold seawater, knee-high, surged forward along the central passageway and cascaded down the stairs into the mess and the galley. With the boat taking more water every second and the stern still dropping, the *Delaware,* with more than three dozen people aboard (roughly half sleeping below), stood in danger of being swamped. Then a fisherman scrambled to the controls and released the groaning winches to slack the cables. The stern surged up out of the waves, the water sluiced off the deck, and the pumps went to work. That was as close a call as NOAA cared to see. A new rule forbid the *Delaware* from towing before a heavy following sea, and when the cruise ended, the boat went to the boatyard, where the thresholds of the rear lab doors were raised to 2 feet.

The boats did not stop fishing rough bottom, however, for the scientists needed to retain fidelity to the survey's randomly selected sampling points. Many fish preferred the shelter of craggy surfaces, and if they were retreating to places like that, NMFS wanted to know. If that meant losing a net now and then, so be it, and damn the expense. A full net rig—net, cables, and doors—ran $13,000. The *Delaware* and the *Albatross* each carried three full sets. That way a cruise could lose two without having to return home. This actually saved money, for with the survey cruises costing $10,000 a day to run, it made no sense to abort one for lack of a $13,000 net. But even with two spares, cruises sometimes had to return to port for a net. At least one cruise had turned home because it lost three entire rigs in succession on the same piece of bottom. As one NMFS staffer described it, "Lost net, cables, and doors; rerigged; returned to the same exact spot and lost another set; rerigged; went back and lost another set; headed home. Three tows, thirty-nine grand, cruise over."

Although Barbosa and Magan and the other fishermen aboard the *Delaware* knew the logic behind this, they understandably disliked the work and the danger it produced. (When the cables started shuddering, any scientific crew would dash for the shelter of the wet lab, while the fishermen stayed out on the fish deck trying to solve the problem.) The NOAA crew didn't like it, either. Sometimes the pilot wandered back and forth over a tow point for an hour or more as the captain watched the bottom scanner for the smoothest piece of bottom arguably on the tow point, and the chief scientist pressed to drop the net smack on the center of the tow point. When the fishing crew finally dropped the net, everyone would wait to see who would be proven right—the scientists knowing that if the net hung, they would probably be idle that shift, the fishermen knowing it might mean several hours of net work out on the deck. The net had hung up and lost a section the day before, and Barbosa and Magan had to sit out on upturned buckets on the rolling, wet deck for a couple of hours using their big plastic netting needles to reconstruct the net's matrix of 3-inch diamonds one knot at a time.

"Well," said Galbraith. "Maybe better luck today. How's the new winch set-up working?"

"Not real well," said Barbosa. "Winch keeps breaking."

"Again?"

"Not today. But you know last night it broke or something. You knew about that. Snapped a gear or something. Then something else little this afternoon. I'm not sure. I think Murph said it happens again we might head home."

"Uh-oh," said Galbraith.

"I wouldn't mind a week home," said Barbosa. "Catch some stripers. Grill 'em."

"Well, me too, actually," said Galbraith, laughing. "You really think the stripers'll be in?"

Five minutes later they had moved from striper fishing to striper cookery—Galbraith had recently bought a new grill and wanted Barbosa's recipes for grilling bass steaks—when suddenly both men stopped talking and looked up at the cables. I hadn't felt it, but we'd snagged. The cables shuddered. Then a cracking sound came from the winch mounted on the roof of the wet lab. We all stepped back into the doorway.

"That doesn't sound good," said Galbraith.

The boat throttled down, which made it rock all the more. Presently, an engineer appeared on the roof, then another engineer, and they looked at the winch for a minute and went back inside. A moment later the phone in the lab rang. Galbraith answered it. It was the bridge. Something had indeed gone amiss with the new gear.

"Looks like a slow shift," he said to Barbosa. Barbosa shrugged.

"I better go to the bridge," said Galbraith, "and radio the *Albatross*."

Everyone dispersed. I descended to the galley and poured some more coffee. From the TV alcove off the mess came the sound of gunfire and shouting; someone had started a video. I stood a moment watching the sea swallow and then release the portholes as the ship rolled back and forth, then went back upstairs.

Off the passageway next to the wet lab was an alcove with a couple of computer terminals. It was unoccupied. I sat down to check my e-mail. The ships sent and received e-mail three times a day, roughly around meal-times. Ostensibly this schedule was to economize on satellite time, but I half-suspected it was designed to prevent crew from spending too much time checking e-mail or surfing the Internet. The schedule, along with our near-complete isolation otherwise, gave the e-mail on these cruises a strangely staggered quality, replacing the medium's normally immediate back-and-forth dynamic with an accelerated version of regular mail's over-lapping delays. (The effect was similar, I suppose, to the several-times-a-day, quick-turnaround London postal service of the early twentieth century.) The pace encouraged you to take a little more time composing, but it also created anxiety if you were waiting to hear anything in particular. Any given load of e-mail could bring news, conversations, or work requests you might rather do without. As Galbraith's boss, Tom Azarovitz, put it, "Used to be when you went to sea, you went to sea. You were gone and the world stopped. Now it doesn't." E-mail indeed proved a mixed blessing on the cruises. It relieved boredom and could salve homesickness, but it could also exacerbate loneliness or worry regarding people or events back on shore.

So it was with my e-mail that morning. I got some much-welcome mail from my luminous sweetheart, Alice, and a short note from my lovely, lively seven-year-old son, but also had to stomach some ugly news from elsewhere in the family and troubling information about a sick friend's terminal illness. There was doubtless even worse news waiting in the papers, and I was glad I couldn't get to them. The world seemed full enough of discord and trouble already.

I logged off, finished my coffee and tossed the cup, and gathered my things to head up to the bridge. The bridge was always the quietest place on board, farthest from the engine and by design a calm, well-ordered place—a good place to escape chaos and, with the big field glasses the crew kindly let me use while there, a good place to watch birds. I'd feel better, I figured, if I went up there and watched the gannets.

On my way up the stairs I ran into Galbraith coming down.

"Hey," he said. "Just the man I'm looking for."

"Yeah?"

"Yeah. The winch is busted. The *Delaware* is returning to port."

"Cruise over?" I asked. I was crushed. Though I'd been homesick only a moment before, the trip suddenly seemed much too short.

"Yes," said Galbraith. "But there's room on the *Albatross* if you and Dan want to go over there. I just checked with Linda; it's fine with her. Jorge'll run you over in the Zodiac if the seas don't get any bigger. But you have to move quick. Can you pack in fifteen minutes?"

"Absolutely."

"Okay then. Meet us at the Zodiacs on the port side in fifteen minutes."

A half-hour later, Doolittle, Barbosa, Magan, and I (Barbosa was going to help out on the *Albatross,* not to get his week home after all) were being lowered over the rail of the *Delaware* in a swinging inflatable boat onto swells that looked a lot bigger once we were bobbing among them, tall enough to block the view of the horizon. Barbosa started the outboard and cast off. The outboard coughed and immediately died, leaving us no way to control our puny boat. We were in the lee of the *Delaware,* but the waves still lifted first the bigger boat and then us, bobbing us up and down in a syncopated sequence that threatened to slap us up against the hull of the now huge *Delaware* just as it came down. I pictured us capsizing, my laptop, with a week's worth of notes, heading to the bottom. But Barbosa, cursing, yanked the crank until it caught, and before the *Delaware* could smack us, we shot off toward the *Albatross,* a hundred yards away. Circling round under the three-story bow to the starboard side, we grabbed and then climbed a ladder hung down the 15 feet or so from the fish deck.

"Welcome!" said Linda Despres, a big smile on her face. "Welcome to the *Albatross.*"

6

The *Albatross* was altogether a different ship, 187 feet to the *Delaware*'s 155, and wider, a full story higher, almost twice the displacement, and blessed with far more space inside and out. The fish deck was much larger, with plenty of room for the fishing crew to work the net and the volunteer scientists to sort the catch, and the boat had two large, open foredecks (the bigger one up in the bow was nicknamed "Steel Beach" for its summertime sunbathing possibilities) that provided high, open platforms for catching some wind or sun and looking for birds and sea mammals. The boat was also more commodious inside. It had two "lounges" off the (much roomier) wet lab for the scientific crew (basically, small rooms with L-shaped benches around a table, but a big improvement over the lack of common space on the *Delaware*); a TV/video room upstairs that was a quiet place to read when no movies were showing; and a computer room just aft of that with four terminals on which to check e-mail or work. The boat had a much nicer balance of enclosure and confinement, the *Delaware* leaning too much toward the latter. The expanse you could find on its open upper decks offered a welcome antidote to the otherwise claustrophobic containment that a boat's small spaces can bring.

Linda Despres, the chief scientist, ran her crew well, though she gave all the credit to them. Along with several NMFS employees of various experience, the scientific crew on that voyage included a teacher pondering a career change, a staffer from the New England Aquarium, a grad student sampling DNA from haddock to see how many distinct populations lived on the Bank and in the Gulf, and another grad student, an Australian named Gavin Begg, who was working on similar issues regarding cod, and whose work on population-boundary issues would later prove helpful to me as I considered various mysteries of Gulf of Maine fish ecology. It was a cheery, rowdy bunch that, something in the manner of a group that has backpacked together for a week, had melded into a good team over the six days of the cruise so far. They were enjoying their work.

Their enthusiasm was fanned without any real apparent effort by Despres, who after more than 1,200 days at sea (this was her 145th scientific cruise, her 60th on the *Albatross*) still loved being on the water. Despres was a natural. While growing up on the coast of Maine, she had spent most of her summers on her dad's 36-foot boat, spearfishing for tuna. "That's when a good day was ten or twelve fish," she told me. "We'd stack 'em like cordwood. Now, a good day is a single fish." She went to Woods Hole for the first time one summer day after her sophomore year in college. She didn't know then the differences among NMFS, the Woods Hole Oceanographic Institution, and the Marine Biological Lab, but she did know Woods Hole was one of the world's premier marine research centers, and she was charmed by the compact village. She looked down the street and thought, "I want to work here."

She figured she'd need a Ph.D. to do so. But in her first year after getting her bachelor's degree in marine biology, when she was working in a grocery store, she saw an ad in the paper for a research-assistant position at the Northeast Fisheries Science Center, applied, and got the job. "Got hired, really," she explained to me that first afternoon, "because of this *one day* I had spent on a Maine Department of Fisheries boat, where I'd counted shrimp larvae in plankton and learned to sort and dissect fish. *That* had been a goofy cruise. The department was so terrified of anyone learning that a department boat had carried a college-age woman on board overnight—terrified my presence would destroy the morals of the crew, I guess—that instead of sailing at 10 P.M. to get to the fishing grounds, as they usually did, they waited until 2 A.M. and left then. I still slept in a bunk as we sailed out—I had my own room, of course—but that way it would not show on the log as an overnight with a woman on board.

"So that was my field experience. When I filled out the application at NMFS, I said that I had sea experience counting and analyzing both fish and plankton populations. I didn't mention it was for only one day."

At the time, in 1973, very few women worked at NMFS in nonclerical positions, and women almost never went to sea on NMFS cruises. For a while, Despres had to work hard to recruit female volunteers for any research cruises she wanted to go on, for she couldn't go unless she found female roommates; the boats needed every stateroom at capacity to compose full crews, and NOAA was not about to allow people of opposite sexes to share a room, even in alternating sleep shifts.* Despres found a way to get to sea often and has ever since. When she sails now, she is invariably chief scientist, responsible for overseeing the two watch chiefs (who direct and work with the two scientific crews), compiling and double-checking the daily log reports, and in general making sure that the data going back to shore and eventually to Steve Murawski are accurately collected and recorded.

Like Barbosa and Magan, Despres had no doubt this cruise's catches paled in comparison to those of previous years.

"There's simply no question," she told me that first afternoon. The *Albatross* had been pulling the same sorts of tows the *Delaware* had, and this hadn't changed that day. Shortly after I boarded the *Albatross,* the net dropped a load composed of one yellowtail flounder and 4 pounds of longhorn sculpin. Another tow that afternoon, which I helped out with so I could finally cut into a gadid, was one of the richest yet—a few dogfish, a basketful of skates, a half-basket of yellowtails, and a dozen or so haddock. At just under 150 pounds of fish, it was the eighth biggest of the fifty loads the *Albatross* had landed so far, including two giant dogfish loads.

---

*Fearful of the complications and jealousies that even the suspicion of sexual relations can create on a lengthy cruise, NOAA still prohibits mixed-sex room arrangements and any sexual relations among crew members, paid or volunteer. This also applies to married couples—the idea being that the knowledge among the rest of the crew that *someone* on board is getting some would be distracting and possibly divisive. Occasionally, of course, the will finds a way.

"It's the worst I've ever seen it," said Despres as she sat in the scientists' lounge preparing the daily report. "Worst anyone on this boat has seen. Ask 'em. [I did; all agreed.] We used to have full checkers several times a day, and in the fishiest sections, every single load. Tow after tow of full nets. This fishery's a mess."

The skimpy loads taken that day—369 pounds in 6 tows, an average of barely 60 pounds each—were the more troubling for having come out of Closed Area 2, one of the three areas on Georges that the New England Fishery Management Council had established as no-fishing zones in 1994. Area 2 straddled the easternmost part of the U.S. section of Georges. (The rich Northeast Peak, as the tip was known, was won in the early 1980s by Canada in a post-Magnuson legal tug-of-war at The Hague.) Area 1, at the western end of Georges, had produced a couple of the larger loads of gadids earlier in the trip (443 pounds of cod one day and 210 pounds of haddock the next; no great shakes, but seed for hope), so Despres was hoping to see some rebound here in Area 1 as well. Fish move around a lot, naturally, and have trouble recognizing the boundaries of the closed areas. But the areas were off limits because they were prime groundfish spawning areas, and with the habitat left undisturbed, one would hope to find here the beginnings of any recovery of Georges' groundfish. It didn't seem to be happening.

I asked Despres what she thought of the regime-shift hypothesis, the fishing-down-the-food-web theory, and the general notion that we had overfished the Gulf and Georges so severely that we had broken something fundamental.

"Well, it worries me," she said, her hand resting on the stack of haul logs documenting the lack of fish. "What worries me most is that there is so little sign that anything is replacing the fish we used to have. So little recruitment, as we call it. No new troops. Dogfish, sure, though even *they're* showing signs of overfishing. But no *gadus* [cod or haddock]. You've seen it. Even when we catch cod, they're all of the same couple of year-classes, three- and four-year-olds. Where are the young of the year?" She held her fingers 6 inches apart. "Where are the year ones and year twos?" she asked, spreading her hands a bit farther. "They're just not out here. I've never seen anything like it."

I reminded her, rhetorically, that this was just the impression from a few days of only one cruise.

"One cruise?" she said, smiling, her eyebrows high. "We're two-thirds of the way through the first half of a year's survey."

"So you're not worried," I asked, smiling too, "about drawing premature conclusions based on what some might call anecdotal impressions?"

"Ah!" she said, raising a hand, her grin growing yet broader. "You forget." She lowered her hand with a smack on the pile of logs next to her at the desk. "I've got the data."

As we headed east toward the end of Georges' long, submerged peninsula that afternoon, the seas turned rough again. The waves smacked impressively against the hull next to my bunk as I lay down for my predinner nap. By the time I woke, the boat was rolling so violently that I had a difficult time climbing down from the bunk without taking a flyer. We pulled in a light load just after dinner, then enjoyed a rolling hiatus before Barbosa and his new crew partner, Tony Veiro, fetched us 400 pounds of dogfish around 10:30 P.M. In between, the scientists relaxed, a few upstairs watching a video, others reading, some checking e-mail, four of them playing cribbage. Finding the main lounge empty, I pushed a couple of sweaters behind me and sat in the booth next to the table to read fishery-science history. This is not an easy thing to do, for fishery science—the science of assessing fisheries and counting the fish in them, as opposed to studying fish (ichthyology or marine biology) or studying the ocean (oceanography)—has been pushed out of the main rooms of ocean studies and into back service halls and utility chambers. As I'd found in the months before the cruise, you cannot easily collect a neat stack of a half-dozen good histories of North American fishery science, read them, and be done with it. You must assemble a jigsaw picture from institutional histories and records, relevant sections of general oceanography histories, review articles here and yonder, and a slim selection of rather dry academic histories that look mainly at the technical challenges fishery science has faced over the years—less a tidy stack of books than a mess of photocopies, overdue library books, and web-page printouts.

In the hours before setting off from Woods Hole, however, I had managed to scare up one of the few works dedicated explicitly to the history of fishery science, an obscure and expensive ($80) volume written by a fishery scientist named Tim Smith, titled *Scaling Fisheries: The Science of Measuring the Effects of Fishing, 1855–1955.* Unlike most fishery scientists who had written on their field's history, Smith had a good grasp of the discipline's large historic trends and a knack for describing them. In his surprisingly fascinating book he pays particular attention to what he calls the central conflict of fishery science, that "between the immediate need to predict catches and the longer term need to understand population and ecological mechanisms that ultimately limit them"—in other words, the conflict between the desire to know *right now* how many fish are in the ocean (and thus how many can be caught without overfishing) and the need to gain a wider, deeper understanding of why fish populations change and how the larger system works to create fluctuations.[8]

Smith (who generously lent me his own copy of the book) had a keen eye for the fault line between these pressures. He worked for the Northeast Fisheries Science Center through the 1980s and 1990s, and though his book goes only to 1955 (probably a wise move, as writing history of contemporary doings is touchy, especially when one of the subjects is your employer), being an NMFS employee doubtless sharpened his perception of a rift that opened in the earliest days of fishery science. This rift was revealed explicitly in one of the first formal stock assessments, in the 1860s, when cod declines devastating Norwegian coastal communities and the banks that financed them led the government to hire oceanographer Georg Ossian Sars to find out what was happening to the fish. Yet the government commissioned Sars not to limn broad ecosystem dynamics but to produce useful answers fast. It was paying him to "examine our cod-fisheries, in order to arrive at practical results that may be useful to our fishermen"—the sooner the better.[9] The big picture could wait.

And the big picture has waited, repeatedly, ever since. Questions about the whys and hows of population dynamics constantly compete for attention with (and usually lose to) demands for the most precise possible immediate head count. The science continually back-burners the pursuit of a comprehensive understanding of biology and ecology that would enable it to predict population changes more accurately and further out in time. Exceptions exist, of course. Even some NMFS scientists find the time and funding to explore the interlaced forces that make or break fish populations—investigations of how trawling affects sea-floor habitat, for instance, or the role that spring storms play in survival of fish larvae. But at NMFS, as elsewhere, the demand for short-term answers has generally dominated the research agenda. In a very real sense, the scientists stay so busy counting fish that they don't know where they're coming from.

It doesn't help that fisheries science programs in the United States have always struggled—and continue to struggle—to get funding for even their most basic, practically oriented research.* As Smith and others have noted, about the only time fisheries researchers receive generous or even

---

*As previously noted, for instance, in 1871, Spencer Baird founded the U.S. Fish Commission (later the Bureau of Fisheries and then NMFS) to conduct broad scientific investigations that would reveal why and how fish populations fluctuated, yet he found within twenty years that he could get money only if he promised to spend most of it on fish hatcheries, the ultimate "skip the science, just fix it now" approach. The commission's research program subsequently became so atrophied that the commission had to hire Bigelow to see what fish were in its most important fishery. Bigelow's work was funded only because the situation was desperate—hatcheries had failed, the Gulf's cod and haddock had taken a dive in the early 1900s, and no one knew why. Even as Bigelow's survey work was nearing fruition (and had already yielded much practical information about the Gulf and its fish populations), the Bureau struggled to get funding for it from Congress. In 1922, Representative Joseph Walsh complained on the House floor that "[the Bureau of Fisheries] has been diverted [into] scientific inquiry . . . [and] fish-hatching operations and replenishing of the supply of fish by these various hatcheries throughout the country has been sacrificed in order that a lot of scientific pamphlets and works might be compiled and distributed." Cutting the research appropriations, he opined, would mean only "the elimination of the employment of some scientific gentlemen who conduct experiments and inquiries that have been made over and over again for the last 25 years." (Both Walsh statements are as quoted from the *Congressional Record* by Hugh Smith in a letter to Henry Bigelow, March 4, 1922; Papers of Henry Bigelow, Harvard University Archives.) There was enough similar sentiment that by the mid-1920s Bigelow often found himself stranded because the commission couldn't replace its aging boats. The desire to conduct broader research—and the need for a boat to do it with—eventually inspired Bigelow to lead the movement to establish the Woods Hole Oceanographic Institution, which he directed for its first decade, 1930–1939. Thus, NMFS lost the services of that particular scientific gentleman.

adequate funding is when fish populations crash, dragging the industry down with them. At these times, of course, the pressure for quick answers grows especially intense, sharpening the tendency to take the close-focus, short-term view.

As I sat reading Smith's book aboard the *Albatross* it struck me that the entire last quarter of the twentieth century was such a time for NMFS and especially for the Northeast Fisheries Science Center, as successive waves of fish depletions had steadily strengthened the call to just count 'em now. The pressure, in fact, had been essentially institutionalized by the 1976 Magnuson Act. "Cooperative management" balancing the economic needs of fishermen with the stock assessments of scientists may have sounded good at the time. But by ignoring the possibility that fishermen might have some well-founded ideas of their own about the health and size of the stocks and giving NMFS the sole, unassisted job of assessing the fish populations *and* of advising the councils on how to achieve desirable catch levels, Magnuson's council system placed NMFS in an unavoidably adversarial position that produced increasingly defensive science. Challenged constantly by fishermen who often saw things differently and in particular by a highly vocal sector of the industry averse to restraint, NMFS had to steer its limited resources into producing the most narrowly focused, ambiguity-free, statistically defensible numbers possible. Magnuson's council process thus transformed NMFS increasingly into a number-crunching machine that elevated math, statistical analysis, and single-species stock assessments far above wider biological inquiries. It also impoverished the industry extension and gear research work the agency had done in the 1950s and 1960s, when biologists, engineers, and fishermen used to team up to develop more efficient, safer, and more effective nets, fish traps, and other equipment. Ironically, both effects further widened the gap between fishermen and scientists and encouraged the sort of math-driven, statistical fishery science that alienated fishermen and excluded their knowledge.

These pressures also helped render NMFS science ever more dry, sterile, and, as John Galbraith put it, indigestible. This dessicative problem had similarly infected many sciences by the digital daze of the late 1990s. Too many scientists, funders, researchers, and journal editors seemed convinced that life, color, opinion, anecdote, or any hint of authorial personality undermined scientific authority. NMFS merely surrendered to a particularly strong strain of this virus.[*]

Part of a general trend in many sciences toward specialization, statistical methods, and a shortage of field studies, this surrender was also an understandable survival response to the constant attacks the service's work faced at council hearings. Nothing resists a beating so well as something dead. So it was with NMFS's assessment science in the post-Magnuson years. The assessment reports grew ever more quantitative, with less conjecture about why fish were found in certain places at certain times and more spreadsheets simply stating where the fish were. The formal reports to the council digitized fishery science down to its most rigid and skeletal form. Both the formal and more informal reports tended to omit anything smelling of blood or opinion. This effect even worked its way back onto the *Albatross* and the *Delaware,* where the daily cruise reports to the center from the chief scientists included less comment and anecdote as the years went on. On the afternoon I talked with Despres in the scientists' lounge, she described going back through cruise reports from the 1950s and 1960s for a data reclamation project and finding, to her simultaneous delight and dismay, a level of anecdote she never saw in more recent cruise reports, even from herself and Galbraith, the two most ebullient, anecdotophilic of the regular chief scientists.

"All sorts of things in them," she said. "Some of it's quite funny—who's been puking over the sides, who's smoking smelly cigars. You get a picture

----

[*]I do not mean to suggest that these trends completely ruled either science in general or fishery science in particular. Fishery science, like many other disciplines of biology and ecology, has its share of good, field-oriented researchers and research programs. In New England, fishery researchers at several major universities (including Dartmouth College and the state universities of Massachusetts, Maine, New Hampshire, and Rhode Island) and at private institutions (notably the Island Institute, the New England Aquarium, and the Gulf of Maine Aquarium, as well as the private institutions in Woods Hole) have conducted numerous field-intensive investigations, sometimes in cooperation with researchers from the Northeast Fisheries Science Center. Some of these projects have involved fishermen in ways ranging from hiring their boats as research vessels to having them help design and conduct entire studies. For the most part, however, few of these efforts have been incorporated in any way into either NMFS's stock assessments or the fishery management council's deliberations.

of real people doing real science. Now we just 'collect data' and sign our names. It doesn't have the personality or interest it had before. There's little reason anyone would want to read it.

"And it's not as if we never see anything remarkable. There's just no place to put it. No sense that anyone's interested. There's no box that says, 'Oh, isn't this interesting.' I remember this one cruise in the Gulf of Maine during the monarch butterfly migration, where millions of butterflies overflew us, and *millions* landed on the ship. Everywhere—rigging, rails, nets, work counters. My boss at the time was chasing them with a big dip net. Where do you put that? Or a rare whale bone that comes up in the net? Maybe in an e-mail. But there's no place for it in the science we're supposed to be doing."

A reasonable person might ask whether any of this mattered. Obviously it mattered to a lot of fishermen, who resented that NMFS's formulas allowed little room for the sort of "anecdotal" knowledge that fishermen possessed in abundance. ("You want to piss off a room full of fishermen," Jay Burnett told me a few months later in this same scientist's lounge, "just say the word 'anecdotal.'") It mattered also to some of the people doing the science—to Galbraith; to Despres and Burnett; to many other scientists inside and outside NMFS. The desiccation of fishery science, in fact, seemed to bother people to a degree directly proportional to their time spent at sea (that is, $I = n/365$, where I is the irritation quotient and n is the number of days spent each year at sea).

But did it really matter to the science and the management? If the center got its numbers right and gave managers the information they needed to manage the fishery properly (whether the managers actually did or not), did it matter that NMFS left out the juicy stuff? Notwithstanding chaos-theory notions about butterfly wings causing hurricanes a hemisphere away, did it really matter that butterflies landed on the *Albatross*?

I thought about all this as I lay in bed that night, still smelling dogfish, still seeing their bloodied skins, but comforted, in a rainy-roof-sound sort of way, by the washy thud the hull produced as it repelled the waves. The boat's now familiar oscillations lulled me, lifting me slowly into weightlessness and then catching me as I fell, compressing me with a gentle, swinging gravity. The motion matched beautifully that of some Bach sonatas for violin and harpsichord I was listening to on my headphones. It was an exquisite, eerie set of works recorded by a pair of players who in their cover photo look almost laughably mismatched. The harpsichordist, angular, gray-eyed, gray-browed, and bookish-looking, sits placidly at his instrument, one hand on the keyboard and one on his knee, smiling shyly and looking slightly embarrassed at his company: a younger, goateed violinist who might have come from the Siberian steppes, standing with his violin cradled warmly in his arms, a steamy, supremely confident look in his dark eyes, and his bow dangling long and straight before him. It was hard to imagine the two having coffee or wine together, and when I'd first seen the cover of the little cassette Alice gave me to take along on the trip, I thought, "Whoa, this should be interesting."

It was. They played together splendidly. The violinist's earthy, sensuous syncopation (a nice change from the metronomic, almost mathematical interpretation many players feel obligated to bring to Bach) alternately hovered above and then dipped down to mesh with the harpsichord's spidery, intricate lacework—rising, floating, and hanging, then dropping with a sweet gravity to merge into music at once rigorous and deliciously rich.

PART III

I

Once when I was on the phone with Dave Goethel, at that point in 1998 when the New England Fishery Management Council was starting to talk seriously about closing the Gulf of Maine cod fishery, I asked him what he would do if he had to stop fishing. He had finished dinner a few minutes earlier, and in the background I could hear his wife, Ellen Diane, and two teenage sons bantering at the dinner table.

"Ellen and I talked about this," he said. "I'm not going to sell the boat. It's paid for. I've been doing it too long, I've got too much into it. Even if I have to fish only one hour of one day a year, I'm going to keep fishing."

If any New England fisherman stood a chance of continuing to fish, it would seem to be Dave Goethel. With a bachelor's degree in biology and more than two decades on the water, he knew intimately not just Ipswich Bay, the rich water that lay off his doorstep, but how to read and negotiate the turbulent regulatory currents that played an equally large role in dictating his fishing possibilities. It helped that Ellen worked, too; she had a thriving, hands-on marine biology education program, "Explore the Ocean World," which she gave under contract to schools. She brought in a fairly regular income roughly equal to a teacher's, which took pressure off the ups and downs of Dave's income. And they helped themselves by being frugal. Back in the late 1980s and early 1990s, when catches of cod and flounder were running high and everyone with a boat and a net was making money—kids barely out of high school owning boats and new pickups and building homes—the Goethels spent their share, taking their boys to Europe and the Soviet Union, but they also saved. They didn't build a big house. Dave didn't get a bigger boat or fancy up the *Ellen Diane*. In a time where everyone wanted more and saw no reason not to have it, the Goethels felt they had enough. So when fish declines and regulations squeezed the margins on fishing, Dave had a well-maintained, debt-free boat, savings to tap, and a head start on finding fish. He thus had a leg up on greener fishermen carrying boat loans and big mortgage notes and also on many of the older, more traditional fishermen who lacked his understanding of the biology and regulations.

Which was probably why he figured he could keep fishing. He was not alone in this thinking. Most New England fishermen suffering through the severe combination of plunging fish populations and rising regulations had similar plans. Fishing doesn't select for easily discouraged people, nor for those who like working inside. Some exercised a stubborn, damn-the-reality determination; others, like Goethel, a more measured resolve. Yet many of the fishermen had gotten out, including quite a few who had said they "didn't know what the hell else" they'd do—who had said, in essence, that they'd stop fishing when someone pulled their nets out of their cold, dead fingers. They'd retired, gone bust, taken jobs. If they lacked a college degree, they banged nails, fixed cars or boats, worked on road crews, detailed cars, or found jobs in fish plants stamping frozen fillets from fish caught in Chile. Those with degrees looked for the rare jobs where they could use their knowledge of fish and boats. They chartered their vessels to scientific researchers. They cleaned up their boats and removed the fishing rigging—the rusty transoms and scarred net spools and scaly fish bins—brushed on a fresh coat of paint, and took tourists to see whales and seals. And while only a handful would deign to work for the National Marine Fisheries Service, some took jobs with state departments of marine resources. "Working for the dark side," one fisherman-turned-regulator had joked. Then, less jocularly, "I go down to the docks now, I feel like a stranger."

So even though Dave Goethel appeared to have good odds of sticking it out fishing, he recognized my question as legitimate. He knew things would get worse before they got better.

"A lot of people who fish," he told me, "think the regulations will get better soon. They're dreaming. If you think that, you should quit. Except those of us who can't see doing anything else," he then added, happily ignoring that this category included nearly every fisherman who had a boat and enough money or credit to buy another day's fuel.

"So I don't know," he said. "I figure I'll keep fishing."

But what if that didn't work? I asked. What if they wouldn't let you fish? What if that one day a year didn't make ends meet? How would he supplement that one hour of fishing?

"There's a number of things I could do," he said. "Work as a research boat. Lobster. I don't know. Gets bad enough, I guess I could always go back to running charters."

At this his wife and sons burst into howls of laughter. "Yeah, right, Dad!" said one of his sons.

"They think that's funny," said Goethel, laughing a bit himself. "They think I'm a cross between Captain Bligh and Blackbeard. Bligh and Blackbeard's Fishing Charters. I didn't always have the most patience with those people. But I could do it if I had to. After fifteen years working for myself, I'm pretty much no good to work for a boss anyway.

"I mean, I could make more on shore with less aggravation. But I don't want a cubicle. After I graduated I did this internship at the New England Aquarium in Boston. Good place. I worked on the top floor. No windows. One day I went into work, it was a sunny winter day, bright out. I came out at five, there was 5 inches of snow on the ground. I'd missed the whole thing. Didn't even know. Didn't have a clue. It wasn't right. So I left after my internship. You fish, you get to watch the weather."

2

Dave Goethel, freshly forty-five when I met him, had seen plenty of weather since that snowy day in Boston and for that matter had seen plenty before then. He'd started fishing seriously when he was twelve, when his dad, a lifelong avid recreational angler, would take him on party-boat outings, driving an hour from their home in Needham, Massachusetts, just southwest of Boston, to Seabrook, New Hampshire, where a small fleet of boats owned by a man named Bill Eastman would head out each summer day to hunt for mackerel, cod, bluefish, and striped bass. The elder Goethel took his son along "for cover," says Goethel, "but I ended up being even more fishing-crazy than he was." Goethel figures he went out forty or fifty days that first year. By the end of the summer, he was helping to bait hooks and clean the boat to reduce his fare. The second summer, at thirteen, he was crewing—baiting hooks, unhooking fish, renting rods and fetching sodas for the sports, scrubbing the boat at the end of the day. He crewed every summer for the rest of his school days, from early June until Labor Day, seven days a week most weeks. "I couldn't get enough. People were paying me to fish? I thought they were crazy."

When he was nineteen, he went to Boston to apply for his charter captain's license from the Coast Guard. You had to have accumulated at least a thousand days on the water just to sit for the test. Goethel presented documentation for well over that. The Coast Guard administrator scratched all the Sundays he had worked the previous seven years because, as Goethel recalls it, "He said no kid should have been working seven days a week all those summers. They weren't used to having nineteen-year-olds take the test. I don't think they wanted me to take it." To the administrator's chagrin, however, Goethel had 1,080 days even after losing all the Sundays.

The test was a full day of written exams that evaluated knowledge of navigation skills and rules; the meanings of all the various buoy, flag, and bell patterns; and fish species, weather, and biology. Goethel finished and turned the test in and sat down to wait for the results. After a while the administrator called him in and told him the commander wanted to talk to him. He showed him into the commander's office. The commander told him he had all but aced the test, answering everything correctly except for two questions in the flag-pattern section. Goethel told him that in that case he had gotten only one answer wrong, because they both knew that one of the flag questions had no correct answer: It was a pattern that didn't mean anything. The commander smiled, then told Goethel that he issued licenses at his discretion and wanted to ask him a few more questions.

"I think he thought maybe I'd cheated somehow. So he asked me five more questions. I got 'em right and he gave me the license."

That summer he captained the fifth boat of the Eastman's fleet—the last to go out each day, the one that sat idle on slow days. He still managed to make some money on his captain's share—half the boat's income after expenses for fuel, bait, and equipment. The second summer, "getting smart," he said, he doubled that take by creating a concession on the boat offering sandwiches, snacks, sodas, and beer and renting rods superior to those Eastman's offered. He split that extra money with the crew. The short New Hampshire coast lacked the cachet, scenery, and sense of isolation of Cape Cod, but the fishing was almost as good. Cod were coming back from the beating the Russians and East Germans and Portuguese had given them, and bluefish were plentiful. (The striped bass, today's staple charter-boat target because of its springy fight, silvery beauty, and tasty flesh, had not yet been rescued from its depleted state of the late 1970s and early 1980s.)

Thus Goethel worked his way through college. He met some commercial fishermen around the docks, and during school breaks, when there was no charter business, he started crewing with them. After earning his degree in marine biology, he took a variety of jobs in his twenties, but he kept running charter boats in the summer and, more and more, worked on commercial boats in fall, winter, and spring. By then he had married Ellen, whom he'd met in college, and they had moved to Hampton, New Hampshire. When he was in his mid-twenties, he and a partner bought a party boat; a couple of years later, he bought a 24-foot gillnetter and started setting gillnets for cod on off-days in the summer and fall.* He was on the water close to two hundred days a year. He grew increasingly absorbed in the art of "thinking like a fish," as the saying goes, reading the currents and the bottom and gauging the water temperature to figure out where the fish would be. He couldn't learn enough about it. On bad-weather days, if he was caught up on his boat maintenance, he would drive to the Agassiz Library at the Museum of Comparative Zoology in Cambridge and read fishery texts and journals to learn more about why fish moved and where they gathered.

"The Europeans," he says, "have always been much sharper on how cod behaved than we have. I even got some good information from some French journals, though I could understand only about three-quarters of it."

Meanwhile, the party-boat business was wearing thin. Goethel took

---

*Gillnets are set into the water and hang curtainlike between buoys; fish swim into them and become trapped, usually at their gills, in the mesh of the netting; the fisherman returns and retrieves the net at the end of the day. Fishing them successfully requires knowing the habitat and travel corridors of the target fish.

some nice groups out but also too many "idiots," as Goethel calls the igno-
rant, drunk, or boorish. People smoked pot in the bathroom or got drunk
and puked on deck, and now and then someone became dangerously surly.
Several times he and his mate had to subdue drunk belligerents, including
one man who pulled a knife on the mate. "We tied him to a chair," says
Goethel.

The outings provided some amusing stories of the dumb-tourist genre
that forms a staple of humor in every town that gets regularly overrun by
tourists. Such stories provide a sort of off-stage revenge that makes it easier to smile at fools while they're there and to laugh at them once they
depart. One story from Cape Cod, for instance, has a summer visitor asking a storekeeper what on earth the locals do when all the elegant people
go home at season's end, leaving only the few thousand locals in their
sleepy towns; the storekeeper answers, "Madam, we fumigate."

Charter-boat captains have their own subgenre of tourist tales.
Goethel likes to tell of the customers—several every season—who, fishing
for stripers or jigging for cod, hook up instead with seals and think they're
fighting huge trophy fish. The seals, cruising the shoals looking for the
same fish the anglers are casting for, grab the anglers' flickering bait and go
on a lightning-fast tear—up to a hundred or more pounds of pure swimming power—that strips all the line off the reel in a few seconds. "And
every single time," recalls Goethel, "the guy fishing is convinced he's catching a monster striper. You can't tell him otherwise. I mean, you've seen this
once, you know instantly it's a seal. Nothing else could take line like that,
except maybe a shark. But these guys start yelling to their friends, 'Whoa!
Look a' this! Big one!' Then the seal takes all the line and breaks it off at the
reel or just explodes the whole rig. They're standing there holding the handle. I saw one guy fall flat on his back when the line snapped off. Another
time I saw four guys hook up with seals all at once. They all thought they
had four record-breaking fish on. They probably still believe it.

"Maybe a half-hour later—maybe—you can explain to them it was a
seal. But at the time, no way. They can't hear it."

Six summers were enough for Goethel. In 1982, he sold his half of the
party-boat business to his partner, sold his gillnetter, and with the fifty
grand he netted made the first payment on a contract to have a 44-foot
trawler built by John Williams of Hall Quarry, Maine, about five hours

north on the coast. The boat would be a Stanley 44 (named after the designer) powered by a Detroit Diesel engine. It was a particularly solid design, in rough terms something like a large lobster boat. Coming from the renowned Williams yard, it would be almost twice as expensive as many boats of similar size. At the time, you could build a 44-footer for around $80,000 or $100,000. But Goethel liked the Stanley 44 and wanted one that would last. He paid $180,000 for it. It has lasted. "It's seen seventeen years of hard use now," he said toward the end of 1999, "with well over 50,000 hours on it, and it's as good as the day it was built. Not many boats you could say that about." He had the boat brought down to his local yard in late winter, finished outfitting it, and was ready to fish by April 1983.

When the *Ellen Diane* first motored out of Hampton Harbor, it joined a fishing fleet that, in response to financial and tax incentives the Magnuson Act and other programs provided, had roughly doubled since Goethel had started fishing a decade before. By the mid-1990s, it would be painful to contemplate programs aimed at putting more fishing power on the water. But in the late 1970s, with the fish rebounding and the Gulf and Georges free of the huge factory trawlers, there seemed unlimited room for the relatively small boats, 40 to 125 feet, that New England fishermen favored.

The incentives were attractive. The standard investment tax credit, for instance, available to anyone but especially appealing to capital-intensive industries like fishing, gave boat owners tax credits for investing in fishing vessels; meanwhile, the fishery-specific Capital Construction Fund program allowed fishing-boat owners to set aside and invest pretax dollars for later use in upgrading or buying fishing boats.

One of the most notorious of the programs updated by Magnuson was the Fishery Vessel Obligation Guarantee Program (or FOG—should have been a warning signal right there), which provided government-guaranteed boat-building loans at lower interest rates and longer payback periods than traditional five-year boat loans. Bank loan officers, finding the risk-return ratio on the government-guaranteed loans attractive, pushed them so hard they inspired a new standard joke in many fishing towns: Open a savings account and the bank would give you either a free toaster or a free fishing boat; and they were all out of toasters. The banks found plenty of takers. In an era of 15 to 18 percent business loans, it was hard to pass up several hundred thousand dollars at 9 percent to build a boat that would fish in what was considered an underfished, rebounding fishery, and the program (along with the other incentives) attracted many investors who had never fished. Many banks pressured both the new-comer-owners and more traditional fishermen to build large, expensive boats. When Dave Goethel tried to use the FOG program to build the *Ellen Diane,* the banker refused him. "We can't do a boat that small," the banker told him. "You want to take a twenty-year, seven-hundred-fifty-thousand-dollar loan and build a ninety-foot boat, come back and see me." Goethel went elsewhere, took a five-year loan, and paid it off, he says, in "just a few days short of four years. I hate debt."

These various incentives, along with the vision of vast waters suppos-edly now lightly fished, produced an explosion in the fleet size. Between 1976 and 1982, the number of boats in the New England groundfishing fleet doubled, from 600 to 1,200 boats.* As half the boats present in 1970 were more than twenty years old, this doubling in the late 1970s greatly modernized the fleet.[1] Because the newer boats were big and carried sophisticated nets and fish-finding technology, the fleet's fishing power grew even faster. Although it took six years for the groundfishing fleet to double in numbers of boats, it doubled its landings of cod, haddock, and yellowtail flounder in only four years. By 1980, the New England fleet was

---

*Most of these 1,200 boats (which do not include lobster boats or other vessels that do not seek groundfish) sailed out of a few relatively large fishing ports. The long-time Massachusetts fishing towns of New Bedford and Gloucester were easily the biggest ports, with several hundred boats each, followed by a handful of medium-sized fishing centers such as Portland and Stonington, Maine; Portsmouth, New Hampshire; and Cape Cod, with several score boats each, and another twenty or thirty ports each harboring no more than a couple dozen small vessels. Of all these towns, only a handful (most notably New Bedford and Gloucester) saw themselves as fishing towns; in the others, the fishing fleets supplemented rather than drove the economies. Maine's lobster industry is an exception. Nearly every Maine coastal town has a thriving lobster fishery, and the lobster boats, smaller, cleaner, and cuter than trawlers or longliners, are universally welcomed both for the lucre they create and for the bloodless, odorless manner in which they evoke the region's fishing heritage.

catching 100,000 metric tons of these species on Georges, in the Gulf, and off the Cape, taking about 50 to 100 percent more than the various stocks of these fish could sustain. Though NMFS survey cruises started to show stocks dropping around 1980, the high catch rates continued until 1983, when they started to decline as fish became harder to find. The slide in catch rates was checked by good reproduction rates in 1985 and 1987, masking the true seriousness of the problem; but as early as 1980, and quite clearly by 1983, the writing was on the wall, though no one yet cared to read it: Just seven years after Magnuson ended what everyone acknowledged had been drastic overfishing by the "floating cities" of foreign vessels, the New England groundfishing fleet was overfishing at rates as high as the foreign trawlers had.[*]

This was far from obvious from the deck of a typical fishing boat, however. Even in 1983, after three years of overfishing, both Georges and

---

[*]This is not to say that the New England fleet was catching as many fish as the foreign fleet had, for it was not—there were not enough fish to allow it, because the ground-fish populations never had a chance to recover to the high levels of the early 1960s. When I say that the domestic fleet was overfishing at a rate as high as the foreign fleet had, I mean that the fleet was exceeding the fishing take that would allow the depleted population to recover by as high a percentage as the foreign fleet had exceeded the fishing take that would have allowed the healthy groundfish population of the early 1960s to sustain itself. In the 1960s, the big-boat fleet had caught about twice as many groundfish as the population could lose without shrinking; by the mid-1980s, the small-boat fleet was catching about twice as many groundfish as the population could lose and still hope to recover. This illustrates one of the painful realities of dealing with an overfished fishery: It is far easier to overfish a depleted stock than it is a healthy stock, because a depleted stock needs to grow rather than just maintain a steady population level. A depleted fishery offers far less slack to both fish and fishermen—they both get squeezed exponentially harder as a stock falls further below its optimum sustained long-term population level. A healthy cod stock, for instance, can sustain fishing levels of roughly a third of its total spawning (adult) population; NMFS estimates the Gulf of Maine cod can sustain an annual catch of almost 10,000 metric tons and still maintain a long-term population of about 30,000 tons of spawning-age fish, and that the Georges Bank stock can maintain a healthy spawning stock of 105,000 metric tons even if fishing boats take about 30,000 tons. In the late 1990s, with those cod populations down to roughly a quarter of their long-term sustainable potentials (and therefore needing very high replacement rates to recover in any reasonable time frame), overfishing meant taking anything beyond about a *tenth* of the populations.

the Gulf still held more cod, haddock, and flounder than had been seen there since the 1960s. If you fished well, you brought home a full hold. It was the era of the 5-ton haul and the 100,000-pound trip, or, for the smaller inshore boats, the 3-ton haul and a deck stacked with boxes full of fish. In the big-boat ports like Gloucester and New Bedford, crew members in their twenties were finishing ten-day trips with many thousands of dollars in their pockets, captains and owners tens of thousands.

Small boats like the *Ellen Diane* did well, too. Like most fishermen, Goethel had many of his best years in the early 1980s and found plenty of reason to be optimistic. His library and boat-deck studies of fish movement, along with a willingness to work endless strings of long days, paid off in catches steady and plentiful enough to pay off both his boat and his house within five years.

Still, like many small-boat fishermen, Goethel fished at smaller volumes and with less intensity than bigger boats did. Boats big enough to go to Georges—big enough to withstand the storms they might be caught in and big enough to hold crew quarters, a shower and a head, and a galley and several days' food—cost around a million dollars to build. Along with monthly loan payments in the five-figure range, they ran up huge crew, fuel, maintenance, and equipment costs. The only way to meet these expenses was to bring in the big catches that an unusually rich and concentrated fishing ground like Georges could offer. If you owned a boat like that, your best bet was to specialize in large quantities of groundfish, and cod in particular, for no other species had their history of abundance and marketability. Once you or your captain got good at catching cod, it made little sense to bother chasing anything else, even if you started to get the uneasy feeling you were overfishing the cod. If you got it in your mind to give groundfish a break and hunt herring or mackerel instead, you would discover you had to spend tens of thousands of dollars to replace your bottom-trawl net and refit the boat with a midwater trawl. Shrimping required yet another type of net and different sorting and storing equipment. You might buy a dredge for scalloping, which could pay well quickly, but generally you stuck to groundfish. And these other fisheries demanded not only new equipment and outlays but new expertise, so even after you refitted your boat to catch, say, shrimp, you still faced the problem of finding them, and it was unsettling to climb a new learning curve with a boat that cost thousands a week to put on the water. So you tended to head for your favorite spots and look for cod and flounder.

Small-boat captains, however, can switch quite nimbly from one type of fishery to another. In fact, they have to. Limited in geographic range by the boat's size and lack of amenities, a small-boat captain must fish and think flexibly and make do with what fish can be found within a few hours' steam. Pushing further means not only discomfort (for most fishing boats under 50 feet lack heads and galleys) but danger, for any fishing vessel under 50 feet or even 60 or 75 feet faces great risk if it finds itself out of reach of harbor in a storm.[*] Sixty to 75 feet is considered the minimum safe size for an offshore boat in New England waters.

Boats smaller than that fit most comfortably in the inshore niche, the 5- to 10-mile-wide strip of shallow bottom (much wider at Nantucket Shoals and a few other places) that slopes out from the New England coast to the 50-fathom line. Though these shoals are intricate and sometimes tricky to work, they offer good habitat that for centuries has supported huge populations of groundfish, particularly cod, vast schools of herring, hake and other species, and mind-boggling numbers of lobster. If fished with some restraint, this inshore fishery is capable of supporting several hundred vessels like the *Ellen Diane* (and several hundred boats working lobster pots). Such boats can do well here. Small enough to work agilely among shoals, lobster pots, and strings of gillnets, they're big enough to tow some gear and hold some fish but small and cheap enough to live on what the inshore fishery offers and switch among the species that move through inshore waters over the year. Now that it's paid for, for instance, the *Ellen Diane* costs about $30,000 to $40,000 a year to run and maintain. Very roughly, a third of that is fuel, a third upkeep and repairs, and a third insurance and mooring costs that Goethel would pay even if the boat sat idle. At that budget, fishing about 250 days annually, Goethel can keep the boat in decent order and outfit the *Ellen Diane* with three different sets of gear that let him pursue about a half-dozen different target species or species groups over the course of the year: a bottom-trawling groundfish

---

[*]The *Andrea Gail,* the swordfish boat sunk by the storm described in Sebastian Junger's *The Perfect Storm,* was 72 feet long; it is one of hundreds of boats over 60 feet that Northwest Atlantic storms have overwhelmed.

net for cod, haddock, pollock, and flounder in late winter, spring, and summer, a raised-footrope groundfish net for herring and whiting during the fall, and a shrimp net for winter.

Catching this variety of fish reliably over a limited geographic area requires a wider, more varied knowledge than does fishing for just one species. You have to know your piece of bottom well and understand why fish congregate where they do. Fishing Ipswich Bay since he was twelve, Goethel had paid a little more attention to such things every year, and by the 1980s, he knew nearly every knob, gully, ridge, slope, tower, and hole. He knew how the water moved and how that affected both the baitfish that formed the middle of the food web and the big fish that he was hunting for. He knew along which fathom-contour lines cod and shrimp and herring and mackerel and flounder collected at what time of year and under what sorts of weather and sea conditions, and he knew how to figure out, most of the time, where they had gone when they weren't where they were supposed to be. In particular, he knew cod. Like many New England fishermen, he had always liked fishing for cod best, both because cod consistently brought good prices and because they were the finest, loveliest fish this water seemed to offer. While not a sentimentalist, neither was Goethel indifferent to the cod's stature along New England shores. It might have been a stretch (if only a small one) to call the cod the fish that changed the world, as one book title has it, but it was no stretch at all to see that cod had everything to do with what New England was and had become, and to grant it a sort of matter-of-fact iconic status—the way, say, the dairy cow ranks in Vermont or Wisconsin, beef cattle do in Texas, or (less majestically) the potato in Idaho.[2]

Unlike the dairy farm business in Vermont, it still paid to know and work cod. Goethel had studied cod intensively both on the water and in the library at the Agassiz Museum, and by the time he started fishing the *Ellen Diane*, he knew them well. He knew when the first cod of the spring appeared just north of Cape Cod; he knew from his friends down in Gloucester and from reading old accounts where they concentrated as they traveled north along the coast; he knew why, depending on temperature, salinity, and current, they collected where they did; and he knew why and when they moved up and down in the water column and why they sought deeper water as summer progressed. He knew how, if not why, they tended to gather around the southeasternmost ridge or knob in an area of high bottom. He knew how long they hung around Ipswich to spawn, and he knew the variables of rainfall, water temperature, wind, and time of year that would drive them off into deeper water or along the coast toward Maine and then out of reach.

And he was lucky enough (or smart enough, since he chose to move there) to live on Ipswich Bay, the most fecund cod ground in all the Gulf.

Rich's *Fishing Grounds of the Gulf of Maine* ably describes Ipswich Bay:

> [It] extends from the north side of Cape Ann about to Portsmouth [New Hampshire] and is resorted to in winter by large schools of cod coming here to spawn. Shore soundings deepen here gradually from the land, reaching 35 to 40 fathoms at 6 or 7 miles out. Within this limit is mainly sand, though rocky patches are numerous between Newburyport [Massachusetts] and Cape Ann. . . .
>
> The principal cod-fishing grounds of Ipswich Bay lie off the northern shore, from Newburyport to the entrance of Portsmouth harbor, 1-1/2 to 5 miles off the land in 12 to 25 fathoms.[3]

A simple map in Rich's book shows this "cod wintering ground," which is just off Goethel's doorstep. This wintering ground, as Bigelow noted in 1953,

> is probably the most important center of production [for cod] for the inner part of the Gulf of Maine north of Cape Ann [even though] this ground is limited to a rather small and well defined area. . . . One consequence of the limited extent of these spawning grounds is that the cod congregate on them at the spawning season in great numbers. During the spring of 1879, for example, when fishing was less intensive than it is at present, and when the cod may have been correspondingly more plentiful, more than 11,000,000 pounds of cod, most spawning fish, were taken on the Ipswich Bay ground alone by local fishermen.[4]

Forty years later, Ipswich was one of the few places in the Gulf where cod collected in concentrations similar to those found on Georges, gathering there in dense numbers beginning in February and staying around to feed and spawn through May and sometimes into June, and supplemented as well in most years by great numbers of cod following baitfish north along the coast as the water warmed. While many similar (though smaller) inshore spawning areas along the coast had been fished out in the mid-twentieth century, never to recover, the cod spawning grounds of Ipswich Bay had survived those first decades of stern-trawler assaults, for reasons that are unclear—some combination of almost unsmashable abundance, local ecology, and, possibly, incoming fish from Georges that brought new recruits around Cape Cod and into the Gulf.

Dave Goethel believes that one reason Ipswich Bay lasted so long is that local fishermen had observed a taboo against fishing at night.

"There was this ethic among everybody that used to fish up here," he said years later, "that you didn't fish at night. It wasn't the only thing like that. They had these . . . well, not quite superstitions, but beliefs and rituals, taboos. Like the dime under the mast or pouring a bottle of whiskey on the net to bless it. That's who told me about the nighttime thing: the guy who made my first net. He'd been a crewman many years, fished everywhere. He told me when I drove off with the net, 'You don't fish the bay at night.' He said, 'You fish the bay every day, you catch fish every day. You fish it at night, you kill it for a week. I hope I don't hear about you fishing it at night.' They didn't know the fish spawned at night or that they were easier to catch then because they rose up a fathom or two and left the protection of the bottom. They just assumed or sensed they were doing some extra damage at night. I mean it was obvious: You fished at night, you caught a lot more fish. You could come in here and work a day—or used to, when you were allowed to—and catch, say, a thousand pounds of cod. But if you worked the same ground at night you might catch four thousand. Which is great, except you'd be overfishing them. So people just didn't do it. I still refuse to do it."

Whether it was the night-fishing taboo or good habitat or infusions of new fish from elsewhere, Ipswich Bay still yielded a profusion of cod when Goethel started fishing it with the *Ellen Diane* in 1983. With stocks at a recent high after recovering from the foreign boats, he found a lot of fish. His fishing soon fell into a comfortable, profitable rhythm: cod in spring and summer, then flounder, whiting, and herring in the fall, shrimp in winter. Once or twice a year he would haul out the boat for maintenance work for a week or so, and he tried each November, in the dead zone between the end of herring and the beginning of shrimp, to take a vacation with the real Ellen Diane and spend some time with his two boys. He paid down his debts. He got involved with some of the New England Fishery Management Council committees and advisory groups, providing a fisherman's perspective on the various plans and amendments and framework adjustments that composed the ever-shifting mosaic of regulation.

During the 1980s, as the fish got a bit harder to catch every year, he came to realize that there were too many boats fishing too many days. He did not think, as some fishermen still insisted even by the century's end, that the Gulf and Georges could take anything we threw at them. He knew that even this fishery had a limit to what it could take and that we were reaching it. And while he had differences with NMFS and thought both the agency's science and the council's regulatory process showed serious disconnections from the way people actually fished, he believed that NMFS had its basic numbers right. He was particularly disturbed by some of the numbers coming off of Georges, where overfishing was at its worst. As he described it, "The council and the industry had the point of view that the fishery was here to take, and they simply weren't interested in restraining themselves. Restraint, frugality, all of those morals and values of the old-time fishermen, they went right out the window in the Reagan years of greed. It was 'Get what you can as fast as you can.' Hasn't changed much since."

When the Conservation Law Foundation suit finally forced the council to close much of Georges to groundfishing in 1994, Goethel winced for his Gloucester friends but cringed to think of the fishing pressure that would come to the Gulf. He and some of his fellow Gulf fishermen suggested that the larger boats be restricted somehow from coming inshore. The council wasn't interested.

"These fisheries get squeezed this hard," says Goethel, "it's like a balloon: You squeeze it in one place, it pops up in another. Every time you pass a new regulation it creates some behavior that creates another problem. Tell guys they can't hunt tuna, more guys go lobstering. Tell them they can't fish Georges, they come inshore here. The only way to end that is to reduce the number of boats. You have to bring the number of boats and the number of fish into equilibrium. Until you do that, you're not solving the problem. You're just delaying."

Delay the council did. Castigated in the most dramatic fashion by the aggrieved fishermen they had just kicked off Georges—entire rooms full of angry fishermen, some of them in tears, begging the council not to send them to ruin—the council did little to restrict fishing in the Gulf and almost nothing to distinguish between big boats and little or to keep the larger boats from moving into the territory traditionally left to smaller ones. The council never went near the discussion that Goethel and many others wanted to prompt regarding how many boats the fishery could tolerate. Instead, it fell back to "democratic," across-the-board measures, such as reducing days at sea for all groundfishing boats and increasing net-mesh sizes—measures that still left the larger boats with far more fishing power without restricting their fishing range.

In the dissolution of traditions and boundaries that both spurred and was sped by the fishing crisis of the 1980s and 1990s, the taboo against night fishing on Ipswich fell. In the early 1990s, even before the council closed Georges, the bay began to fill with nighttime lights as big boats turned from the failing Georges Bank fishery to work the bay. It must have seemed too easy, working inshore just a few hours from Gloucester and New Bedford and Portland, over much softer bottom than on Georges, sweeping up the cod as they rose to feed.

"They came from everywhere," said Goethel, "and they went at it every night. We used to call the bay the Thousand Points of Light, there were so many boats. I wrote the council saying, 'You can't let this go on. You're destroying the fishery.' But the big-boat guys that can go around the clock, they don't want to see that happen."

For Goethel and many others, everything changed when the big boats left Georges and came inshore. Along with increasing the pressure on their home-ground fishery, the shift illustrated starkly the most alarming facts about the New England fishery. Anyone who knew anything recognized that 90-foot boats were working inside only because they had already fished out one of the world's most productive fisheries. That they were allowed to come inshore made clear that the council was unlikely to stop them from doing the same to the Gulf.

By the mid-1990s, the New England fleet was catching Gulf of Maine cod at more than twice the sustainable rate, according to NMFS, and had cut the cod population by half since 1990, reducing it to about a third of its sustainable level. Even so, a good fisherman like Goethel could still catch decent loads on most of his mingy eighty-eight-day annual ground-fish allotment. With that and the shrimp season and the flounder and herring and whiting, Goethel continued to do okay. He wasn't doing nearly as well as he had back in the 1980s. Not hardly. But he was in the black, even if by slim margins. By the time I first fished with him, in late 1998, he was frustrated and perturbed but essentially comfortable. It looked as if he could tough it out.

3

"Let's see if we can get out of here without hitting anything," said Goethel. "I can't see a thing." In foggy, predawn darkness, the way out of Hampton Harbor is not obvious. Our immediate navigational challenge was avoiding the concrete pillars holding up the highway bridge spanning the harbor's outlet. Goethel flipped on the spotlight atop the wheelhouse, but the beam just produced a wall of glare, so he turned it off. The darkness that replaced the glare was so complete he turned the light on again just in time to illuminate, about 15 feet in front of us, one of the bridge's 6-foot-thick pillars. Goethel spun the wheel right. Good news: The *Ellen Diane* could slalom. We went right, then left, and slipped between the pillars into the jettied channel leading to Ipswich Bay. We couldn't see the jetties. The buoys marking the channel emerged out of the fog at the last second, usually more or less in front of us. We ran over some of them; they scraped along the side of the boat and popped up behind. Finally, we passed the last buoy and entered deeper water. Dave pushed the throttle forward. His crewman, Charlie, a quiet, kindly man in his fifties, went to the small bunk area in the fo'c'sle to catch a nap. It would be a half-hour or so before we got ready to set.

We were hunting for "herrin'," as New England fishermen call herring, pronouncing it heron, and also hoping to catch some whiting, or silver hake. Midwater fish. Goethel had started fishing for them in the late 1980s after he recognized that while most Americans didn't care to eat herring (though the Brits like them in "kippered" form), American lobsters do, which means there is a steady, reliable demand for herring by the thousands of New England boats that bait lobster pots. Ipswich has always been rich in herring. So Goethel, like quite a few other fishermen, took to spending two to four months a year pulling herring out of the water and selling them (indirectly, via the fish market he uses) to lobster boat captains, who put them in lobster traps and put them back into the water.

First, of course, he had to find the herring. Fish finding is a strange art, and I was curious about how Goethel, whom several people had named as being particularly proficient, practiced his. He said, sure, come along, explaining that he was actually better at finding cod, but since NMFS hardly ever let him fish for cod anymore, he'd have to show me how he hunted herring.

Finding fish, he had told me, is a matter of knowing where you're supposed to find fish (that is, where the topography and current and bottom type are of the kind your target fish prefer), where people have traditionally found fish before, where you yourself have found fish before, and a good guess as to where you'll find fish this time. Fish tend to gather along breaks in topography—drop-offs, fences, ledges, holes. When you fish for herring, for instance, you tend to fish at 30 to 40 fathoms, dragging the net just above the bottom—though within a given year, month, or even day, variations in water temperature, current, or other factors might move the fish up or down in the water column or to a deeper or shallower fathom contour. Likewise with cod. You tended to find them between 20 and 60 fathoms, shallower in spring and deeper as the water warmed, and you would often have your best luck hitting the southeasternmost aspect of a ledge, knob, tower, or other lump or ridge in the sea bottom, or perhaps the southeasternmost knob in a series of knobs. They tended to gather there first, and if you found the right spot you might find a dense school of cod only a half-mile from where another boat was scooping up hardly anything.

Some of the factors and rules of thumb you could write down, memorize, pass along, look up again. To this end, most good fishermen keep a log. Goethel keeps his in a thick, weather-beaten, 6-by-9–inch spiral notebook in which for each outing he records the date; time of day; location;

weather; sea direction, size, and condition; time of set and length and depth of tow and its yield; and any oddball thing that might seem to pertain: whether there had been birds around, or whales, or signs of baitfish near the surface; whether the area was getting worked hard by other boats.

Most of this information is tangible and measurable—data that can be tracked and recorded by a computer. The act of writing it down helps you memorize it, burns it into the brain at least subliminally for recall or unconscious consideration later, and writing it down also offers the opportunity to draw or emphasize connections a computer might be indifferent to—that, say, an unusually huge haul pulled off the southwestern face of a ridge (a place you'd tried before with little luck) came during a waxing-moon ebb tide on the afternoon of a calm day that followed a three-day blow.

Like a good journal, good fishing logs are faithfully kept even if rarely consulted—and so when consulted, capable of yielding great insight. Goethel writes in his virtually every day, but doesn't read it often. However, if he strikes out for a couple of days or gets results that don't jibe with his remembered experience—if the combination of knowledge and instinct in his head doesn't bring him fish or if a chance set brings him fish where he wouldn't have expected them—he will look up the log entries for similar dates in years past. What he finds may not give him a definitive answer to whatever puzzle he's trying to solve, for, as he puts it, "The fish never do the same thing two years in a row." On the ocean, as with rivers and one's home of youth, you can never return to the same place. Yet his log sometimes reminds him of something he tried that worked or some conclusion he drew or the relation between a combination of weather events and time and some unusual fish behavior, or it simply reminds him that he has not seen everything yet, so new things might be worth trying.

There was other stuff too, he said, that you picked up sooner or later and just internalized. Big storms along the coast, for instance, often drive baitfish temporarily into deeper water, sending the big fish out after them. Storms offshore might drive baitfish in, bringing cod with them. Heavy rains inland, meanwhile, send cooling flows out of rivers, changing the temperature and salinity close to shore and either attracting fish or forcing them away. Any given year might see cod tending to gather in tight groups, meaning you needed to find them and fish short tows to do well, or it might see cod collecting more loosely, rewarding a strategy of long tows taken over big stretches of habitat.

Good fishermen intuitively mesh all this information into what they call "fishing smarts" or a "sixth sense" or "thinking like a fish," though it might more properly be considered thinking like an ocean: an untraceable calculus that combines articulated, conscious knowledge about depths, location, weather, water temperature, time of year, and recent history with a hunch based on cumulative experience. Good fishermen find fish, in short, the same way a good pitcher chooses a pitch or a good parent the right words for persuading a reluctant child to put his shoes on and get out the door for the school bus: by confidently, calmly applying a mixture of transferable knowledge and untransferable instinct.

This was the sort of explanation numerous fishermen gave me when I asked them how they found fish. It was all correct and it all applied, but in the end it left me unsatisfied. It didn't explain how a fisherman found fish any more than a good batter's explanation of how he keeps his hands back and tries to pick up the pattern made by the spinning seams explains how he hits a curve ball. Something was missing: the part where the bat hit the ball.

I wanted to see bat hit ball. And I wondered what sort of fish finder Goethel was. Was he a pitch-charting plotter or a see-it-and-hit-it type? I figured there were good fishermen of both sorts—the Ted Williams type who charts every at-bat, studies pitchers' pitch sequences, and always stepped into the box looking for a particular pitch in a particular spot; and the Mickey Mantle type, who just went up and hit the thing.[*]

Being guardedly enamored of high technology myself, I also wondered whether Goethel had adopted much of the high-tech navigational and fish-finding gear that had hit the fishing world over the past couple of decades. Like many arts and sciences in the 1990s, fish finding was increasingly practiced by people hoping to substitute technology for experience,

---

[*] Once when Mantle homered deep into the seats in an All-Star game, he returned to the dugout to find Williams, who was among Mantle's teammates on the American League All-Star team that year, awaiting him on the dugout steps. Williams eagerly asked Mantle what pitch he had hit. Mantle shrugged and said he had no idea. Williams was aghast.

knowledge, and cultivated intuition. Sonic fish-finding equipment had been around since the 1980s, showing anyone with ten minutes' schooling in reading the screen whether the boat was over fish. More recently, global positioning system (GPS) navigational satellite technology had come along to tell even the most compass-blind boaters their locations with up to 10-foot accuracy. As one *Albatross* fisherman put it, exaggerating only slightly, the GPS units were so accurate that "you could drop a sneaker overboard, come back next day, and fetch it up with a net." The same software could autopilot your boat, guiding it along a particular course, even one quite complicated, and could also record a course so that you could later replicate any or all of it. This made it easy to return precisely to a productive spot—and fish it bare if you wanted to. Along with improved bottom-imaging sonar and bouncier roller gear that could handle rough bottom, these navigational systems also allowed trawlers to safely work rocky areas that had previously posed too great a threat to their gear. "This stuff has turned the ocean into a glass table," one fisherman told me. We were sitting at his kitchen table. He grabbed the big pepper mill and pulled it toward him. "The stuff's so good you can find a tower like this, which would be completely surrounded by cod—cod just about clinging to it—and which before you would have steered clear of for fear you'd lose your net, and you can fish it so closely, going around and around"—he spun his pointed finger slowly around the pepper mill tower—"that you can pick virtually every last fish off the thing."

Along with bringing fishing pressure to habitat that had formerly been too risky to fish (and thus eliminating many small but vital de facto habitat refuges), all this technology radically shortened the learning curve for new fishermen.

"It cut way down on how long it takes to fish competently," said Goethel. "You used to serve apprenticeships of maybe five to seven years before you knew your way around the grounds—not just how to navigate, find your way out and back home again, but how to find fish and catch them without tearing your gear up. Now you can take someone who's pretty bright and in six months have that person fishing almost as well as a lot of people who've been doing it ten or twenty years. Not as well as the best, but well. Only on the tough days will they show they're green. I think for a while, when things were good, that brought more people in than might otherwise have come in. And now, with so few days, it's made a joke of the idea that the lousiest fishermen will drop out and just leave the best ones. We get so few days now and are allowed so few fish per day, any idiot with a GPS and a good net can come out here and catch his limit. You don't catch much extra for being good. You don't get weeded out for being bad. That's what this great technology has done for this fishery."

Goethel was not a Luddite, however. The *Ellen Diane* sported a half-dozen little computer monitors (rather modest technology by 1998 standards) clustered in front of the aging, duct-taped, split-vinyl captain's chair: a couple of 10-to-12–inch screens on the shelf in front of the wheel, 3 or 4 smaller ones hanging on brackets from the ceiling. Most of them duplicated others. Two sonic fish finders, one about ten years old, one about two, showed the same information at different scales and at a definition crude compared to recent models; they showed only what was just before, under, and behind the boat but could not look to either side or far ahead. A digital Loran readout gave longitude and latitude in degrees and minutes but no seconds; two radars at different scales provided some collision prevention in fog or dark. He also had a GPS unit he was trying out. It was a puny affair compared to the showpiece I had gazed at on the *Delaware*. It showed our location not on a detailed, screen version of the familiar white-and-aqua NOAA chart, but on a black screen of small scale that merely showed fathom-contour lines at 10-fathom increments. This let you avoid any big obstacles and gave you a sense of the general terrain, but it hardly painted the picture that a more elaborate system would.

Goethel was unsure whether he liked it.

"I'm not sure if I'll keep it or not," he said. "I'm not really all that comfortable with all this stuff. I grew up at a time where you navigated with a compass and a watch. I just want a compass, a radar, and a Loran, maybe a fish finder to confirm I'm on fish. Computers—forget it. Computers and I don't get along. They malfunction when I come in the room. At home, I can walk into the room when my wife is working on the computer, and it'll mess up. I tell her, 'Wait a minute. I'll leave; it'll fix itself,' and usually it does. Sometimes when that doesn't work I offer to go get a hammer and see what I can do to straighten the thing out. She always declines.

"Computers. Don't care for 'em."

The way he said it reminded me of a story I heard, supposedly true, of a Vermont farmer, eighty-something years old, who was taken by family to New York City. He had never been out of New England. When he returned, someone asked him what he thought of Manhattan. He said, in those lovely flat vowels of the old Vermonter, "T'ain't necessary."

"There's no doubt this stuff makes it easier to fish," said Goethel. "It makes it harder to go broke. But it won't give you the more subtle knowledge of where the fish will be when and how they move around. That just comes with experience. And not everybody gets those things even if they've been out a while. I mean, there are guys who fish for twenty years who don't pick that stuff up. It's something you either absorb or you don't. Like feeling the net."

"You mean when it snags?" I asked. I was remembering how I'd missed it when the net hung so badly on the *Delaware* that spring.

"Not just when it snags. When it bumps or scrapes things or rolls over different bottom. You learn to feel little vibrations coming up the net cables and through the boat. You sense them and know what the net's doing. You get that little tremor, you've bumped the edge of a ledge. You get a bump, you've run right over some ledge. If it pulls a second, you may have torn a few diamonds out of the mesh. You feel a tremor, it went over a little ledge. You get that bumpity-bump, you bounced over a boulder pile. That's stuff you can't teach. It has to come from experience. Some people never learn it. They'll just suddenly hang up and tear the net or even lose it and say they never felt a thing. I don't know if they're not paying attention or what."

I had not known Goethel very long at this point, and though I had heard similar things from others before, I wasn't sure I believed him. He was talking about vibrations that were traveling up 150 to 300 feet of cable through turbulent water and then through the structure of a boat that was being propelled by prop across the heaving surface of the sea by a diesel engine that ran rough enough to shake coffee out of a cup.

On a subsequent, warmer trip, however, I saw what he meant. Goethel, sitting on an upside-down bucket on deck, talking as we towed for flounder, several times stopped midsentence and looked back at the net cables, his attention drawn to them by vibrations too subtle for me to notice. Later, I talked to yet more fishermen who told me they felt such tremors and signals, feeling the bottom through the net cables. Still later, I did some reading that suggested that even finer, more subtle things could be sensed tactilely.

Indeed, my reading suggested that while I almost surely missed that net-to-cable feedback because I'm a landlubber, I may have missed it because I wasn't sitting down. For Goethel's description of drawing intricate information from the net's transmitted vibrations parallels closely the descriptions by adventurer David Lewis of how the great Polynesian ocean canoeists navigated. Though they lacked compasses, maps, charts, or anything else we would recognize as navigational technology, these sailors reliably and regularly negotiated scores or hundreds of miles of unmarked open ocean with precision, even under cloudy skies offering no stars or sun to navigate by. They found their way, Lewis tells us, by feeling changes in the water, most notably and sensitively with their testicles. Seated in loincloths on the thwarts or bottoms of their outrigger canoes, these supposedly unsophisticated men of the sea could sense, in a manner hard to articulate but that carried to them definite meaning (a manner quite close to truly visceral, one might say), information embedded in the shape, size, and rhythm of swells and waves that revealed to them unambiguously their exact location on the broad Pacific. Jonathan Raban describes Lewis's findings in his own excellent *Passage to Juneau:*

> Whatever the fickle gusts of the moment, the prevailing seasonal wind was registered in the stubborn movement of the sea. Swell continues for many days, and sometimes thousands of miles, after the wind that first raised it has blown itself out. Islands, because they deflect the direction of swell, can be "felt" from a great distance by a sensitive pilot. As the depth of the sea decreases, the swell steepens, warning of imminent landfall.
>
> Sailing by swell entailed an intense concentration on the character of the sea itself. Wave shape was everything. A single wave is likely to be molded by several forces: the local wind; a dominant, underlying swell; and, often, a weaker swell coming from a third direction. Early navigators had to be in communion with every lift of the bow as the sea swept under the hull in order to sense each component in the wave and deduce from them the existence of unseen masses of land. . . .
>
> So did Lewis's Polynesian friends feel their way across the humpbacked ocean. On these voyages, Lewis—a vastly experienced small-boat sailor—often found himself totally disori-

ented, as the wind changed direction, the sea got up, and the underlying swells became confused or imperceptible. Yet his guides could sense a regular grain in the roughest, most disorderly sea. Time and again they'd sail through fifty or more miles of murky overcast, without sight of the sun, and make a perfect landfall at—in one instance—a narrow passage between islands, breaking into sudden visibility less than two miles off.

I never summoned the courage to ask Dave Goethel if he sailed by the balls. Yet he did seem to be one of those fishermen who sailed, as it were, more by the seat of his pants than by his dashboard instruments.

At 5:40, the light outside just strong enough to show it would be a grim, gray day, Goethel slowed to make the first set. Charlie came up from below, and both men pulled on their orange rain gear and knee boots and went to the rear deck. I stood in the doorway to the cabin, trying to stay out of the way and looking over my shoulder every few minutes to peek out the front windshield and make sure no oil tankers or fishing boats had materialized out of the thinning fog. It took Goethel and Charlie about ten minutes to get the net out, first pulling the cod end off the spool that spanned the deck overhead, then feeding it by hand into the water, helping it over the stern as those first few yards went over the back and got pulled behind by the boat's slow progress, and watching the line of floats along the top cord to make sure the net paid out smoothly. When the cod end was out, roiling the gray water, Goethel came forward to the winch controls and set the winches spinning to pay out the cables. When the net was all the way off the spool, he stopped the winch. He and Charlie went to the boat's rear corners, attached the heavy trawl doors to the net, and unhooked them from their racks. Then Goethel set the winch going again. The net pulled the doors back into the water. They disappeared. Weighing several hundred pounds and curved ever so slightly like wings, they would simultaneously pull the net toward the bottom and, their wing shapes pushed outward laterally as they moved forward into the water, spread the mouth of the net wide so it could catch fish. The long net would open up behind like a huge windsock, its top cord held high by the floats, its bottom cord pulled low by its heavy metal rollers.

They had done all this work on a deck that was rolling significantly by my standards, though neither man ever so much as stutter-stepped or raised a hand to balance. The seas were not big—maybe 6 feet—but 6-foot seas on a 44-foot boat are much bigger than 6-foot seas on a 185-foot boat. Watching from the doorway, short on sleep and long on gut-eating coffee, and exposed a bit too much, I was beginning to realize, to the smell of diesel coming up from the engine compartment, I was feeling unwell. When the two fishermen returned to the cabin—Goethel to his captain's chair to glance at the radar, Charlie below for another nap—I crossed to the back end of the deck in the rain and fed the fish.

Afterward, I stood a minute and looked out over the gray water. I hoped to see whales, but the lumpy sea didn't show much from my low perspective, and it seemed a day so ugly that even whales would not come out. Only September, and already the air carried a wet chill. I promised myself I'd get out here when things were really rough, say February, but even as I made the promise, I knew I probably couldn't hack it and would find some reason not to get out then, when a day like this would probably be the very best you could hope for, only not even that good, for it would be about 20 degrees instead of 45.

Back inside, in the warm diesel air, I asked Goethel why he chose to make his first set at this particular spot.

He leaned back in his chair and said, "Well, in this case it's not that hard. We're about the right depth for herrin'. We've been catching 'em around this 30-fathom line all month. Plus," he said, pointing at the sonar, "Mr. Fish Finder here says we're over a pretty good school of something, and I'm figuring it's probably herrin'."

"What else might it be?" I asked.

"Could be dogfish," he said. "Sometimes I think dogfish are God's revenge on fishermen. If it's dogfish, it's going to be a long morning."

PART IV

I

The days of scrotal navigation are over, supplanted by the invention of the compass. In their accounts of how they find fish, however, all but the most techno-dependent fishermen describe using information that, though difficult to fit into any regimented database, bears on how fish populations move, gather, disperse, and wax and wane: on population dynamics. Many Gulf of Maine fishermen speak of resident versus school cod, for instance: the smaller, darker, resident cod (also called "ground" cod) sticking roughly to an inshore area for many months at a time or even year-round and moving offshore, if at all, only in the warmest-water days of summer and fall; the bigger, more silvery, more mobile and transient school cod arriving to feed on schools of baitfish that come inshore only in spring and summer. Experienced fishermen distinguish these cod by appearance. The resident cod have relatively large heads in relation to their bodies (possibly a sign of being underfed), while the school cods' heads look smaller and better proportioned on their fuller, sleeker bodies. Some fishermen also distinguish between red cod and green cod. No one knows for certain whether resident and school cod are two distinct populations within the larger Gulf of Maine, or whether (as Bigelow suspected) they represent cod at two different periods of their life or yearly cycles, so that "a cod that is a ground [or resident] fish this month may start on its travels next, turning brighter and becoming more shapely as it goes, either from a change of diet, from a change of surroundings, or from more active exercise."[1]

In a long passage in *Fishes* describing the wanderings of Atlantic cod populations, Bigelow notes that a lack of tagging studies had left it a mystery, as of 1953, how much or little the Gulf's cod populations moved, whether some moved widely and some stayed put, and whether the distinction between resident and school cod overlapped or even duplicated another traditional distinction many fishermen made, that between relatively shallow-spawning "local" inshore fish and the lesser-known, presumably deeper spawning offshore stocks. Did all the different groups mix together over the course of their lives as one large population? Did some or all of them remain apart as distinct or semidistinct breeding populations?

Although such issues of stock structure strongly influence population dynamics, these questions never made it to the forefront of NMFS's assessments of New England's cod, either in Bigelow's time or since. Rather, NMFS has used since before Bigelow's time a two-stock picture that viewed all northern New England cod as either Gulf of Maine cod or Georges Bank cod. While this broad, simple geographic division ignored distinctions such as those between inshore and offshore cod and resident and school cod, it had served the agency well, allowing it (as Steve Murawski could so effectively document) to track and predict population changes and fishing effects with great accuracy. Many NMFS biologists find finer-scale considerations of stock structure intriguing, but the increasing emphasis in fishery science on statistical modeling, the "count 'em now, damn it" pressures described earlier, and the general effectiveness of NMFS's Gulf-versus-Georges stock picture, have discouraged the use of other stock distinctions that were less obvious and less well-documented.

Yet while NMFS was not making it a high priority to weigh more heavily these fish movements and additional stock distinctions, Gulf fishermen held them central to their conception of cod and how to fish for them. Most Gulf fishermen saw cod as living in a different biological world than that represented in NMFS's assessments and stock divisions. In the days before he stopped fishing groundfish because there were so few, Willie Spear, for instance, a renowned fisherman out of Yarmouth, Maine (just north of Portland), planned his fishing calendar not around the existence of Gulf of Maine or Georges cod, but around the movements of the resident and school cod that NMFS's assessments essentially ignored.

"We'd catch pollock till December," he told me, "then we'd move into the shoal water and fish the resident cod. They were there all year. Market size, but not real big. They didn't look anything like the cod that came and went. They were all scarred up and discolored, dark, whereas the school cod were white-bellied and clean-looking. The school cod moved all the time. The old-timers would show us these, like, little lice, mites, some kind of parasite, on the bellies of the school cod. Said that's what kept them clean. That was another thing you'd look for. The school cod tended to be bigger fish. A lot of them fifty pound or more. We'd start catching them around Easter. Easter would come the first Sunday after the first full moon after the spring equinox, so we'd start on the first full moon of March. We'd do what you'd call 'fish up to the western': Fish over to the Isle of Shoals and further out, meet those school cod coming up, and then follow them up the coast."

Dave Goethel used a similar approach, though unlike Spear, Goethel had a particularly dense spawning population of adult fish right in the Ipswich Bay winter spawning grounds. He could work them from February into April, when he would turn his attention to the bigger, school cod coming up the 50-fathom line from north of Cape Cod.

Along with distinguishing between resident and school cod, another distinction drawn by fishermen but not used officially in NMFS's stock assessments is between inshore-spawning cod and apparently wider-ranging offshore-spawning stocks. Some fishermen use this distinction synonymously with the one between resident and school cod, others separately. Many believe the legitimacy of the inshore–offshore division is proven by the disappearance in the mid-twentieth century of most of the inshore cod from many areas and their failure to return even when offshore cod were booming. Almost all the Gulf's inshore areas hold far fewer cod and other fish than they once did, yet no one really understands why. The answer would be nice to know, as reversing any correctable causes might make the Gulf a far more productive cod ground, particularly for the smaller boats that work such waters.

The disappearance of the inshore populations has largely escaped official scientific scrutiny. NMFS, the state agencies, and the rest of the scientific community have known about it but for various reasons have not made it a top research priority. (Most of the inshore water is state water, so NMFS understandably puts its priorities elsewhere, and Maine, the state with the most affected coastline, had not seen fit to look deeply into this failure of the inshore fishery, perhaps partly because the question of whether pollution from the politically powerful paper industry plays a role makes it a sensitive question.)

One curious fisherman-scientist, however, Ted Ames of Stonington,

Maine, conducted a study in the mid-1990s that sought to document the old inshore spawning grounds and puzzle out the causes for their demise.* Ames, from a long line of fishermen on Vinalhaven, an island in Penobscot Bay, had earned a master's degree in biochemistry with a thesis on marine bacteria in the early 1970s and then gotten a fair way through a Ph.D. program in oceanography at the University of Orono that he did not quite finish because he kept getting up from his desk to go fishing. His spawning-grounds study, a mix of oral history and heads-up scientific thought and analysis, found some illuminating answers regarding the disappearance of the old stocks and also raised some intriguing questions.[2]

Ames's idea was simple: Ask several dozen of the best veteran and retired cod fishermen where and how they used to catch cod in inshore waters, and from their answers create a picture of how cod concentrated along the coast and when they disappeared. From the memories of these older fishermen, Ames collected a wealth of data that were on the verge of being lost. For starters, he found that inshore spawning grounds along almost the entire Gulf coast had once produced huge numbers of fish, whereas by the time of his study, in the mid-1990s, only Ipswich and Massachusetts Bay regularly produced significant numbers. This in itself wasn't news; everyone knew that fishermen used to catch far more gadids close to shore than they had since around the mid-twentieth century, and it was assumed that some of those fish had been spawning. No one, however, had explained why the spawning populations never reestablished themselves once fishing pressure eased. One hypothesis was that the fish were not returning or could not spawn successfully because of pollution from Maine's paper mills or increased rainwater runoff from the state's ever-more-paved landscape. But as Ames pointed out, these runoff problems had never reached many bays farther north along the Maine coast that nevertheless remained bereft of cod. Something else was preventing a rebound.

---

*The study was sponsored by the Island Institute of Rockland, Maine, a nonprofit organization whose mission of preserving Maine's coastal culture had led it to address the fisheries crisis and, in particular, to try and build better dialogue among fishermen, scientists, and regulators. Ames's was the first of several Island Institute projects devoted to that end.

The stories he collected suggested an answer: that these inshore cod spawned in fairly distinct "runs," much like those of salmon, that mixed with the larger Gulf of Maine cod population for much of the year but at spawning time split off and converged on their places of birth. As described by the fishermen Ames interviewed, the coastal spawning grounds tended to be gravel-lined depressions, troughs, or other dips in the bottom, 10 to 50 fathoms deep and located near the mouths of rivers. Local spawning stocks would arrive suddenly in the bays in spring or early summer, concentrate in huge numbers for several days or weeks to spawn, and leave. In some places they stayed so briefly and in so small an area that they would go undetected year after year. They also escaped heavy fishing pressure, even though they were close to shore, because spawning cod are not as tasty as nonspawners, nor do they salt well, so demand for them was poor until the 1930s and 1940s, when the less-discerning frozen-fish market emerged. And even when they were discovered and fished before that time, the hook-and-line or gillnet fishermen weren't capable of taking them all.

When the bottom-scraping otter trawl and the frozen-food plant became common in the 1930s and 1940s, however, these inshore cod runs began to disappear. Ames heard accounts of abundant local spawning stocks in bay after bay that were hit unprecedentedly hard with new otter trawls that were capable, on these relatively smooth bottoms, of scooping up nearly every fish. The spawning run would come, word would get out, and the local fishermen (and their neighbors down the coast, if word traveled fast enough) would fish intensively, the boats wing to wing, until they wiped out the stock. It all happened, on a local level, almost faster than the fishermen themselves could comprehend.

Once fished out, the stocks never returned; they vanished like lights going out on a big wall map. This failure to return led Ames (and many others reading the accounts) to conclude that the inshore fish exercised a salmonlike natal fidelity in their spawning habits, refusing to spawn anywhere except in their birthplaces—and thus not likely to move up and repopulate a bay that had been fished out. In some cases, clean, productive bays that had been fished out were ignored by still-spawning fish in adjacent bays for several years in a row. Neither were these fished-out shore areas reseeded by eggs floating in from elsewhere, for, as Ames notes, the

Gulf's current pattern prevents eggs laid offshore from finding their way into the bays. With no new in-migration and no egg drift, any bay that got fished out stayed that way.

Some of this picture is necessarily conjecture, and Ames presents his conclusions regarding natal fidelity with the requisite *suggests* and *indicates* rather than *shows* or *proves*. With no present population to study, there is no easy way to confirm whether the bay-spawning cod were genetically distinct local spawning stocks; doing so would require some serious digging through archived scale or otolith samples to see if enough genetic material from vanished bay-spawning cod could be analyzed to draw any sort of conclusion. Yet absent such proof, Ames's notion of an entire coastline of locally spawning stocks, like many other valuable theories, seems to solve a fundamental puzzle: why inshore cod have failed to rebound the way offshore cod did in the 1970s when fishing eased.

Ames's work, vetted by some of Maine's leading fisheries ecologists and brought to the attention of the scientific and fishery communities by the Island Institute, gained a quick credibility in that state and was translated into policy with rare speed and directness. To the chagrin of some of Ames's fellow small-boat fishermen, the Maine Department of Marine Fisheries, which regulates all fishing within 3 miles of shore along the state's coast (as do similar agencies in other states), in 1998 cited his study in banning all groundfishing for cod in those state waters during May, June, and July, when any remaining inshore cod presumably were spawning (and when the school cod were also moving through).

Beyond Maine, the work sent subtler ripples. The study gave strong credence to fishermen's insistence that their "anecdotal" knowledge could help manage a fishery if it was collected with appropriate rigor and judgment. It also raised some interesting scientific questions: Do offshore cod populations exercise natal fidelity of the sort the extirpated inshore stocks seemed to heed? Do most or even all cod spawn in very specific locations, move on to live their adult lives, and then return to their home grounds to spawn? Were all the cod spawned on, say, Cultivator Shoals, driven to spawn only there—and if all the Cultivator Shoals cod were caught, would any other cod use that ground? The tendency of eggs laid on Georges to drift many miles before hatching into larvae that then drifted many miles before becoming fish seemed to argue against strict site fidelity in open waters; but perhaps the larvae, having drifted and then swum down to a sheltered gravel bed somewhere together, would then remain faithful to that bed, or perhaps a fish's DNA coding somehow drew it back to the site of its parents' spawning even though it had never been there—a possibility seemingly incredible but with ample precedent in the ever-more-fantastic findings about natal fidelity and migration behaviors.

In any of these scenarios, a steady fishing-out of distinct offshore spawning stocks or substocks by increasingly effective gear might explain the continued lack of recruitment in Georges Bank and the Gulf of Maine. Had we been eliminating discrete, open-ocean, site-specific "spawning runs" one by one, slowly eradicating the spawning populations of the Gulf and Georges? If so, or if that were a strong possibility, what were the implications for managing a rebound? Should we have instituted spawning-season closures or gear restrictions on Georges and in the Gulf long ago? Should we be managing those spawning stocks individually instead of as part of the much larger populations of Georges and the Gulf?

Perhaps most troubling, why hadn't we been asking these questions before?

2

In the fall of 1998, about three days into what was looking to be a slow fortnight aboard the *Albatross* searching for cod in the Gulf of Maine, I asked Jay Burnett who the all-time worst-behaved volunteer scientist had been.

"That would be the writer," he told me, laughing. He related the tale: A writer who wanted to see how NMFS counted fish went on a research cruise on Georges in February. A huge storm kicked up when the boat was at its farthest point from port. With nowhere to run, they stayed and tried to keep working. Seas ran to 30 feet. The boat plunged over the crests and into the troughs so violently that as Burnett lay in his bunk—an upper bunk—the boat repeatedly tossed him up and down. The bunk would lift and then suddenly fall out from under him, leaving him airborne to be smacked down by the ceiling, which normally was about a foot above his nose. Even some of the boat crew got sick. Instead of getting ill and morose, however (the usual presentation of seasickness), the writer became increasingly agitated, approaching hysteria. He paced and talked frantically, screaming periodically about the boat sinking. Eventually, he said he believed the boat was cursed: At his departure his young son, angry at him for leaving for two weeks, had told him he hoped the boat would sink. "And now it's sinking!" the writer yelled.

The boat was not sinking. But the writer was terrifying himself and unsettling everyone else. He kept screaming, "I must talk to my son!"

"When we threatened to restrain him," said Burnett, "he calmed down a *little*. I don't think that guy's been back on a boat."

Our own cruise, in contrast, was encountering calm seas and few fish, leaving plenty of time for storytelling. Jay told me about the scientist who reached without looking into the checker and was stung by a sting-ray, went into shock and, heavily sedated and only semiconscious by the time a Coast Guard cutter came to rescue him, had to be half-rolled and half-flung from the rocking deck of one boat to the other; the Coast Guard crew just barely caught him, pulling him on board by his hair and his clothes. I also heard of the *Albatross* pulling up a piece of a Navy plane (the Navy met them at the dock and quickly took the piece away). I shared my own, smaller store of stories, one about a Cape fisherman who had pulled up television sets still in their cartons (apparently washed off a freighter in a storm); and one about a Gulf dragger fishing alone whose sleeve got caught in the net as he reeled it in—thank goodness it wasn't going *out*—and was wrapped around the spool of his own net. He spun round and round, encased in the net, for almost an hour before a nearby fellow fisherman saw the boat and figured out what was going on, got to it, jumped aboard, and extracted him.

Working so close to land, we could often see the shore. The sight of the crenulated gray-brown horizon made the cruise feel at once less committed and more closely relevant than the Georges cruise. Less removed, as well; walking through the TV lounge, you ran the risk of encountering the five o'clock news. On the other hand, there was a rumor we might dock in Portland, and the prospect of a beer was comforting.

The only thing we caught much of those first few days was dogfish. No car-sized loads this time, but in haul after haul, amid a scattering of hake, lobsters, and flatfish, dogs composed the biggest part of the catch: 825 pounds on the very first tow in Cape Cod Bay; then 140 pounds, 5 pounds, 45 pounds, 135 pounds, and a disheartening 325 the following day. They were disgusting as ever. Cutting them open to see what they'd eaten and whether they had young was my least favorite part of processing the catch. Pushing a knife tip through the tough skin into the abdomen required a pretty good shove. Then you had to saw forward 10 inches or so to expose everything, and the skin put up a gritty, ripping resistance no matter how many times I stoned the blade. The stomach rolled out, often full of water; I'd slice it open and water would gush out, bringing half-digested herring or mackerel, which would compromise fish sandwiches for some time to come. By this time, the worst would be over unless it was a female with young: the finger-sized pups (dogfish, being sharks, give birth to fully formed little dogfish) came out looking grotesquely fetal; tiny little dogfish, fins and all (though no spikes yet) covered with glutinous mucus, disturbingly babylike and sharklike at once.

Hardly a cod to be found. We were working our way north from Cape Cod Bay, zigzagging back and forth across the 50-fathom line and working stations both inshore and offshore along the Massachusetts, New Hampshire, and southern Maine coasts. This water—Goethel's water, much of it—typically held a lot of cod in spring and early summer, but by now the fish had moved off into deeper water and, as both fishermen and scientists put it, "dispersed," meaning no one knew where they had gone. The survey's previous leg, which had done most of the stations in the Gulf's interior, hadn't found many cod either, a mere 355 pounds in 45 tows. Our first dozen hauls fetched up less than 50 pounds. With Ames's paper in my head, I wondered, as we sailed along scooping down into the shallows and depths and not finding anything, just where the fish had dispersed to. Did they scatter through the deeps to mouth blindly the dark ocean floor in search of mollusks and starfish? Did they let the Gulf's gyre carry them out through the Great South Channel to Georges? Did they swim upcurrent, heading east toward the Bay of Fundy? Did they sink into the great holes, Wilkinson and Jordan and Georges Basins, in huge, undiscovered concentrations? Did they wander round willy-nilly, ending up in almost any old place that was comfortable? Or were they beginning to respond to strangely familiar, irresistible signals steering them toward the gravelly swales and rocky ledges where they had been born?

I had gone on this second cruise partly to see if we would find any cod for Dave Goethel to catch the next spring (things weren't looking too promising for Dave so far), to reflect on the question of where the Gulf's fish went, and to talk to Jay Burnett. A few months earlier, about the time I was absorbing Ted Ames's monograph, Burnett had begun a project that resembled Ames's in some rough and interesting ways and seemed to signal a move by the Northeast Fisheries Science Center to reach out to fishermen and make use of their knowledge. Earlier that year, the Science Center's new director, Mike Sissenwine, had called Burnett into his office and asked him if he'd like to work with some New England fishermen to see what research and assessment issues they might help NMFS with. A year or two earlier that conversation would have been unthinkable. But much had changed. Ames's paper had shown up, for one thing. And the same National Research Council report that had found the Northeast Fisheries Science Center's science sound had faulted the center for its distant relations with fishermen and urged the NEFSC to "improve relationships and collaborations between NMFS and harvesters by providing, for example, an opportunity to involve harvesters in the stock assessment process and using harvesters to collect and assess disaggregated catch-per-unit-effort data."[3] So, in rather graceless, unimaginative form, was made official the growing clamor—barely heard two years earlier, even from fishermen, but rising steadily as 1998 wore on—for cooperative research between the NEFSC and the industry.

Sissenwine, an accomplished and highly intelligent scientist who had taken over at the center earlier that year when the previous director stepped up to Washington, saw he had to get something going. "I'm not going to send my assessment team out on fishing boats just so they can get to know fishermen," he told me that summer. "They've got better things to do. But there are some projects we can take on that can improve the science and answer this call for better relations."

Thus his query to Burnett. Burnett answered with the mix of enthusiasm and reluctance of one taking on a mission worthy but possibly doomed. "I told Mike," he related later in the relatively safe confines of the *Albatross*, "that I didn't want to do it if it was just going to be a pretty show to make the fishermen feel like we were listening to them. I said I'd do it only if we could produce some good science."

Sissenwine agreed there was no sense in a feel-good exercise; they'd go up the coast, talk to some fishermen, and see if some issue of substance offered itself. Sissenwine related that his own handful of conversations with fishermen so far had suggested at least one ripe prospect: Perhaps some of the more experienced captains could help the center refine its knowledge of how the Gulf of Maine cod moved through Massachusetts and Ipswich Bays and along the Maine coast in the spring and early summer. If the fishermen who pursued these fish could help the Science Center sharpen its picture of how the schools moved north, the center could better advise the New England Fishery Management Council on how to protect those fish, perhaps using "rolling closures" to protect the main concentration of fish while leaving other areas open.

Burnett recognized that he had been chosen for this work because he was both of assessment and not of assessment; that is, he knew assessment science, having worked in assessment for three years in the mid-1980s (and being married to someone who worked there now), but because he no longer worked there (he now worked in the Age and Growth Division, where he helped track age structure and growth dynamics), he would presumably be less tainted and prejudiced by the bad blood between fishermen and the assessment team and less offended by the criticisms the fishermen would level at the assessment division. As anyone who had been to a council meeting knew, fishermen often said unflattering things about the assessment team. Sometimes the ire washed back the other way. One of assessment's more prominent members once said at a public meeting that he didn't care if *all* the fishermen went out of business if that's what it took to get the fish back. Fishermen remember that sort of thing.

As articulate and knowledgeable and unflappable as was, say, Steve Murawski (and though Murawski would never say something as impolitic as his colleague had), Sissenwine recognized that someone who did not work in assessment would probably have an easier time establishing trust with fishermen. Perhaps he thought it would help, too, that Burnett had worked as a bricklayer for nine years (before a back injury sent him back to school) and understood the perspective of people who finished every workday deeply tired—the iron willfulness that marked the most enduring of them, the resentment they felt regarding people in softer jobs who seemed not to appreciate their work. Burnett had also spent considerable time on the water (another trait almost no one in assessment could claim) and was one of the few NMFS scientists who looked as comfortable on a boat as in an office. (Actually, he looked more comfortable. I once found him sleeping on the wooden table in the *Albatross'* scientist lounge. "I've gotten some of my best sleep on that table," he said.) He was probably the NMFS biologist a group of beleaguered fishermen would be most likely to relate to. He was easy to talk to, constitutionally straightforward, and averse to pretense. And he was genuinely interested in what the fishermen had to say. "He was the first NMFS person in years," Dave Goethel told me later, "to come up and really listen to fishermen."

Over the summer and fall of 1998, with the help of New England Fishery Management Council member John Williamson, a couple of fishing organizations, and the Island Institute (the Rockland, Maine, outfit that had published Ted Ames's paper), Burnett and NMFS anthropologist Patricia Clay met with several dozen of the more experienced captains (including Dave Goethel) along the Maine, New Hampshire, and Massachusetts coasts. The meetings, as Burnett later described them, quickly took on a distinct pattern.

"We'd go in," he told me one day as he sat on the high table in the *Albatross'* scientist's lounge swinging his legs, "and for the first hour, maybe two, the fishermen would just vent. They'd just about scream at us. They'd tell us how sick they were of the assessments and how they wished we'd make up our minds and how sick they were of the regulations changing all the time. I can't blame them. These guys are just snowed with paperwork. Sometimes I think that's part of the effort-reduction strategy: Bury them so deep in paperwork they give up and quit fishing.

"Anyway. They'd bitch an hour or two, and we'd sit and nod and ask some questions, but mainly just get yelled at. Then sometime about halfway through the day they'd get it out of their systems, and they would see we weren't leaving or trying to sell them something, and we'd get to what we came there for, which, to start with, was to talk about the timing of the fish coming up the coast. So after that first round of meetings we had two issues to go back and talk about. One was the movement of these fish. The other was the possibility of a study fleet—a bunch of fishing boats that could help us with things like tagging studies or otolith samples or bycatch studies, so we could get a finer-scale look at some of those issues than we can with our surveys." There was even a funding possibility for such a project through the recent appropriation of $5 million, from a bill pushed through the U.S. Congress by Massachusetts' powerful congressional delegation, that was to be used, in proportions and a manner left rather hazy, for a combination of economic relief to New England fishermen and research on the fish stocks.

After those initial meetings, Burnett went back to several groups to get details on how the cod moved up the coast. To make sure he would not generate a false consensus on how the fish moved, he interviewed each fisherman individually—actually took them one at a time to a separate room, cop-show style—and rolled out a big chart of the Gulf and asked them where the cod gathered when and how they moved. The fishermen all painted the same picture. They described the movement of a large body of cod that started just north of the Cape's northern tip in late winter and slid west and then north, roughly along the 50-fathom line, through Massachusetts and Ipswich Bay and on along the Maine coast. Craig Pendleton, a Portland-area fisherman who ran an umbrella organization of fishermen, fishing industry, and environmental groups called the Northwest Atlantic Marine Alliance (NAMA) and who attended some of the meetings, was frankly envious of Burnett's getting the benefit of all that knowledge: "I threw a hundred-dollar bill down, said I get first dibs on that chart when Jay's finished. Holy smoke. Sit down with a chart and have all those guys tell you where they find fish? I'd pay a hundred bucks for that, easy."

The picture that emerged was consistent and emphatic. "It was like it was written in granite among these guys," said Burnett. "They'd been fishing this way for years. The fish come up to Wildcat Knoll [an underwater knob a few miles north of the tip of Cape Cod] about February, then move north through Mass and Ipswich Bays and on up. One guy said his father taught him always get down to Boon Island [an underwater knoll in upper Ipswich Bay, several hours' south for a Portland boat] on St. Patrick's Day to meet the cod as they come inshore. 'Everybody else would be at the bars,' he said. 'My dad would be out there making money.' It was old knowledge. Anyway, these guys wondered if those fish came from the south, maybe down in Great South Channel or from Georges. They figured if they did, maybe taking those fish as they came up into the Gulf wouldn't actually hurt the Gulf of Maine population, since they weren't really Gulf fish." Most of the fishermen identified these fish as school cod on the move rather than resident fish, and some wondered if the traveling cod, which seemed to come up from Georges, were either part of the Georges stock or some separate stock NMFS wasn't tracking—a notion that would explain, among other things, why many fishermen had continued to find more fish among this annual movement than they could reconcile with NMFS's Gulf assessments. Dave Goethel, for instance, said his fairly good luck catching cod during the mid-1990s had not matched the increasingly dismal NMFS Gulf assessments.

"We've long known that when the cod leave Georges in the winter, they disperse in almost every direction," Goethel told me. "Some move west to the Great South Channel, others west and north into the Gulf. A certain number move over and hit the Cape around the beginning of February and then travel on up and around into the Gulf. You can set your calendar by it."

After listening to other Gulf fishermen talk to Burnett, Craig Pendleton, too, believed that NMFS had misclassified or was missing a major piece of the cod-stock picture. "NMFS has drawn this line at forty-two twenty [that is, 42 degrees, 20 minutes, NMFS's regulatory border between Georges and Gulf stocks] and decided because Gulf cod are beat down worse than Georges cod you can catch only two hundred pounds a day above the line and thousands of pounds below it. But you look at the pattern these guys were describing to Jay, and that line just makes no sense. Because these schooling fish move across it every year."

These ideas intrigued Burnett, who was among many who thought that NMFS's stock division between the Gulf and Georges might be oversimplified. He knew from Bigelow, for example, that early tagging studies had shown that a subpopulation of cod congregating each winter to spawn on Nantucket Shoals (the vast area of relatively shallow water spreading south from the Cape Cod area and separated from the western end of Georges Bank by the Great South Channel) migrated after spawning, though the studies showed more movement west than north. And while Bigelow had thought it likely that most of the Gulf and Georges school cod (which he figured was the majority of the fish) did not wander far in their lifetimes, he noted that this wasn't known with any certainty, and that thorough studies in Norway had shown that some schools of cod regularly travel great distances. At the same time, there seemed plenty of evidence from tagging and other studies, including NMFS's own surveys, that fish spawning in the western Georges/Great South Channel area might be a population separate from the cod that spawned over Georges' eastern half. The Canadian Department of Fisheries viewed Georges cod populations this way and had been pushing NMFS for some time to do the same.

Given all that, it didn't seem far-fetched that part of this western Georges stock might curve north around the Cape and up the route the fishermen described. A gusher of plankton leaves the Gulf through the Great South Channel every spring, attracting millions of baitfish; it would make perfect sense for a cod shaking off the covers and looking around for a meal after spawning to work its way right up the Channel and into the Gulf.

In the science of counting fish, perhaps no question is more fundamental than how fish stocks are structured, that is, what distinctions of geography, behavior, genetics, or life history separate one more or less distinct, independently reproducing population of fish from another. A stock is virtually by definition the ideal unit by which to count, assess, and manage fish populations. In an assessment scientist or fishery manager's perfect world, one would be able to count and manage each stock separately; that would make it possible to always know how each stock was doing and to ensure that no stock got overfished (thus decreasing population size and diversity) or underfished (which would deny fishermen income and everyone else a good food source).

In the real world, however, scientists and managers must compromise this ideal to accommodate various pesky constraints, such as the difficulty of identifying and tracking each fish stock separately, the tendency of most stocks to mix with other stocks of its own species as well as other species, the highly complex interactions of fish, ocean, and fishing boats, and the usual limitations of time and money. With cod, for instance, the cod's wide travels make it difficult to clearly identify and separately track any but the most clearly defined stock, and the cod's habit of mixing with other groundfish, such as haddock, pollock, and flounder, makes it next to impossible to regulate fishing pressure on them (and also on those other species) separately. That's why cod are managed, in general, as a part of the "multispecies groundfish complex."

Given these realities, an optimal assessment and management strategy—one that is realistically ideal—would operate within the most unavoidable of the constraints while recognizing and accounting for the most important stock divisions and each stock's size, dynamics, and requirements. In other words, you would not necessarily manage each stock separately, but you would accommodate each stock's basic nature and essential needs. You would know what separate stocks were out there and how each was doing, and you'd regulate fishing pressure and human-caused influences so that no stock got over- or underexploited.

You could not achieve this optimal practice if you had misread or ignored crucial stock distinctions. Bungle your stock i.d. and you'd likely screw up both assessment and management. You might count and regulate fishing on only a portion of some stock, leaving the rest of that stock untracked and unprotected so that it might get depleted (or do quite well) without you knowing it; or you might inadvertently treat as one stock several populations that should be tended separately—an error that would mask the effects your management had on the individual stocks, some of which might be doing fine, others of which might be getting hammered. And in any of these cases, you wouldn't know any bad news until it was too late, if at all.

That last scenario—failing to identify and properly manage separate substocks—apparently wiped out Ames's bay-specific spawning cod populations, for instance, and it's instructive to think for a minute how a better picture of stock structure then might have changed things. At the time Ames describes those stocks being wiped out, in the 1930s and 1940s, our fishery science and management made only the crudest distinctions about stock identification. No one suspected that inshore cod might reproduce in bay-specific populations. Cod in the Gulf were, as now, considered to be just Gulf cod. But let's pretend it's 1930, that those populations are still intact, and that we know what we today know (or think we know) about stock structure's importance. To simplify the exercise, let's further pretend that for some reason (superstition, lousy boats, a fuel shortage, sea monsters, or U-boats offshore) fishing occurs only inshore, so we need manage only a coastal fishery. If we believe that the cod in each bay comprise a stock distinct from those in other bays and that reproduces separately, we will prevent the individual bay stocks from being overfished, for logic dictates that if we wipe them out, they are unlikely to return. If, on the other hand, we have ample evidence that the bay cod populations are merely the fringe of a much larger Gulf of Maine cod stock from which they are essentially indistinguishable (as was the unspoken and perhaps unconscious assumption), we would have little reason to restrain our fishing. With so many fish, why conserve? (For the moment, we'll set aside worries about biodiversity.) Under these assumptions, it would make perfect sense, from a management perspective, to take most of the fish that came inshore each summer, since we could reasonably believe we were taking only a small, easily replaced fraction of a huge stock.

The picture of stock structure used by fishery managers thus carries profound implications both for the fish and those who hunt them. Fishery scientists increasingly recognized this beginning in the 1960s and 1970s, when studies of troubled species both marine and terrestrial demonstrated the critical role played by the divisions and links between and within populations. The decimation of both Pacific and Atlantic salmon and the discovery that salmon were divided into river-specific spawning stocks showed the importance of recognizing and managing for separately reproducing subpopulations within a species. Similarly, the fragmentation of terrestrial habitat revealed how vulnerable wildlife populations become as their genetic and geographic isolation increases, particularly if their numbers or habitat are shrinking. These findings were greatly enriched by the identification of several increasingly accurate and useful markers—various types of genetic coding, and protein and chemical analyses—that could supplement the morphological and life history tools, such as size, color, growth rate, and feeding and spawning habits, traditionally used to distinguish between different populations.

Wildlife management, however, has often lagged these scientific developments. Policy customarily trails science somewhat, of course, and fisheries management has perhaps trailed further because it is so difficult to parse divisions between huge populations of highly mobile animals living in opaque habitats. That's one reason (along with bureaucratic inertia) the Northeast Fisheries Science Center still uses the same basic stock divisions for cod that it used when Bigelow first sailed. In the official eyes and spreadsheets of the NEFSC—and despite that many of the Science Center biologists recognize these subtler issues—a cod pulled out of northern New England waters is either a Georges Bank cod or a Gulf of Maine cod, period.

Strangely, NMFS's two-stock picture grew mainly out of administrative convenience. In the prewar years, most boats that worked Georges unloaded at Gloucester, New Bedford, or other Massachusetts ports, while most boats working the Gulf (that is, anywhere north of Georges) unloaded at Provincetown, Portsmouth, Portland, or other upper New England ports convenient to the Gulf. The Bureau of Fisheries divided its fishery management and (eventually) assessment regime similarly because doing so made it easy to match assessment and management measures with landings data from those port areas. The agency stayed with this division because Bigelow's work and most subsequent science showed that the single most obvious and important groundfish habitat boundary in these waters was indeed that between Georges and the Gulf. Although some water and fish mixed along the Bank's northern edge and at the channels at either end (particularly the Great South Channel, which is off Georges' western edge), the opposite current gyres of the two areas, along with the steep drops around Georges, appeared to make them separate ecosystems, each with its own distinct population of groundfish.

A century of management and almost four decades of close counting seem to confirm the accuracy and value of this two-stock picture. Even today, no one questions this most basic division. Even many scientists within NMFS, however, recognize that this stock picture ignores some of the stock distinctions that, as described earlier, fishermen have long drawn, such as those between school and resident cod and inshore and offshore. NMFS's Gulf/Georges picture also ignores the opinions of many scientists, including those with the Canadian Department of Oceans and Fisheries, that Georges Bank cod should be seen and managed as two stocks, an east and a west.

Evidence for these additional, finer distinctions (much of it in anecdotal form from fishermen but some in more traditional scientific studies) has existed for decades and grew toward the century's end. In the 1990s, for instance, studies using new genetic-marker techniques suggested that several cod populations in European and Canadian waters that had long been treated as single stocks contained genetically distinct subpopulations that might warrant separate management attention.[4] Similar work as well as tagging studies have confirmed and particularized fishermen's distinctions between inshore and offshore cod.

Nonetheless, NMFS continued to stick with its two-stock division out of a combination of arguably healthy scientific conservatism (if it ain't broke, don't fix it); the conviction (also reasonably defensible on scientific grounds) that the extent and nature of these finer divisions were not clear enough to compel a change in how NMFS counted its fish; the increasing predominance of statistical considerations and methods over biological and ecological, as described earlier; and a combination of practical constraints and what might be called scientificobureaucratic inertia: that is, the resistance to reworking the stock divisions because doing so would not only compromise the value of the long-running survey data but create immense amounts of work recasting and reinterpreting those data to apply them to whatever new stock picture the Science Center might adopt. Resistance was also warranted by the difficulty of deciding which of the additional possible stock or substock divisions to include in any revision: Should NMFS track and manage separately for both inshore versus offshore *and* resident versus school populations—a pair of distinctions that might sometimes overlap, merge, or conflict with each other? How would it track these divisions? Among fish the industry caught, for instance, how would NMFS know which ones were resident and which school, or which inshore and which offshore? How should it handle these subsets within the two or three main geographic stock areas? Ponder those problems for a while and you begin to understand the Science Center's reluctance to change its much simpler Gulf/Georges picture. That reluctance becomes all the more understandable when you consider that the center's two-stock picture had produced reliable, verifiable assessments for decades and predicted with remarkable accuracy what would happen under various fishing levels, including the wild, unintentional experiment in overfishing of the 1980s and early 1990s. Everything had played out pretty much the way NMFS had said it would, so why change? If the stock picture was fundamentally off, surely some error would have shown itself.

If NMFS had a good argument for holding on to its Georges/Gulf cod stock division, others could cite good reasons to question that stock picture. And the hypothesis that the Gulf fishermen were presenting to Burnett—that the school cod constituted a third stock that annually crossed the line between the Georges and Gulf and therefore might not be accurately counted or regulated under the two-stock picture—was particularly intriguing. It was obviously still more hypothesis than tested theory, much less established fact. And Burnett wasn't sure that proving this third stock existed would change things much for the fishermen anyway, at least in the short term. "It wouldn't necessarily mean they could catch more of these fish just because they came from western Georges," he said. "It's not as if Georges is overloaded with fish."

It seemed a hypothesis worth exploring, though, particularly because one of the newer biological techniques, called otolith elemental analysis, held promise for testing it. "That's the big bang in this to me," Burnett said. "You've got this new way to test this and maybe answer the question definitively that doesn't have to rely on the fishermen's accounts. You take otoliths from these different groups of fish and analyze them. If they're different stocks, they should have different chemical profiles in their otoliths."

Otoliths, also called "ear stones," are small, ovoid discs of bonelike material that float in tiny compartments in the center of the skulls of many fish species, including all gadids. Most fish that have them have three pair, the largest of which are called the sagitta. In a full-grown cod, the sagitta are roughly the size and shape of lima beans. What fish use them for, no one really knows—navigation and balance are the leading bets. Biologists, however, use them as age meters, for otoliths are laid down in seasonal surges and slowdowns that create rings that, like those in trees, can reveal a fish's age. Generally, you can't simply cut one in half and see the rings. You have to bake the otolith at 525 degrees Fahrenheit for three to six minutes, slice it in cross-section lengthwise, then peer at it under a 15-power glass and a good light. Amid the yellowed milkiness of the cooked otolith you will see distinctly darker brown rings—the annuli, one per year.

Working in the Science Center's Age and Growth Division, Burnett had spent a lot of time with otoliths. He could also put his hands on thousands of otoliths—no shortage of fodder—for the scientific crews of the *Albatross* and *Delaware* had removed and saved the otoliths from a huge percentage of the gadids they had netted over the previous thirty-five years. Tens of thousands of the bony discs, each pair still in its 1.5-by-3–inch manila envelope inscribed with information indicating exactly where and when the fish was caught, were stuffed into the little cottage that housed the Age and Growth Division in Woods Hole. They filled all the closets and lined the shelves in halls and offices and conference rooms. They were stacked amid the books in Burnett's tiny cubicle overlooking the water. If this otolith elemental analysis business was all it was cracked up to be, Jay Burnett was sitting on a gold mine of possible stock identification data.

Understandably, he found rather exciting the prospect that his tiny little friends might reveal critical information about something as big as stock structure. I say *might* because elemental analysis employs finicky, expensive techniques and had an uneven if promising history. Elemental analysis (sometimes called microconstituent analysis, elemental chemical analysis, or some hybrid of these terms) had first been developed in the late 1960s. It was largely set aside as a stock structure tool in favor of genetic methods in the 1970s before being revived and more intensively developed in the 1990s as the genetic approach showed unexpected limitations. Elemental analysis of otoliths was based on the notion that a fish's path through life—the temperature and chemical make-up of water it lived in and moved through, the food it ate, the environmental stresses it endured—created in its growing otoliths a distinctive mixture of the chemical elements it consumed, absorbed, or internally produced. Roughly 99 percent of any otolith would be calcium carbonate and protein, an otolith's main materials. The other 1 percent, however, would hold a unique (unique, that is, to all fish sharing a life history and habitat) blend of some or all of forty elements, ranging from aluminum and arsenic to titanium and zinc, that the fish's genetic and physiological experience would have etched in the unalterable tablet of its otolith. Fish in a given stock would share an elemental profile distinct from that of fish in a different stock. If you could identify a stock's elemental markers, you could differentiate fish from that stock from fish from others, even when the stocks mixed.

Elegant science in theory. In practice, it had proven challenging. Reliable results required that you have many otoliths, treat them well, choose the most proper and revealing of the four analysis methods available, and know what to look for.* You apparently needed some good luck as well, for by 1998 the technique had yielded mixed results. Some studies of multiple stocks had revealed a dozen or more elements that could be productively compared, while other studies had yielded only two or three, and only a handful of studies had produced results that could be consistently duplicated. (Reproducibility of results, of course, is a major criterion of any experiment or analytical technique.)

Nonetheless, a few studies, including one of cod off eastern Canada, had produced fairly convincing two-step stock analyses, first identifying distinct, detailed elemental profiles in different groups of spawning cod, and then, after the cod had left the spawning areas and mixed in the open ocean, successfully identifying individuals from those spawning groups among fish caught in the mixed schools.

As Burnett explained to me on the *Albatross*, he hoped in the months ahead to replicate, at least roughly, the results this Canadian team had achieved: He planned to send otoliths from both Gulf of Maine and Georges Bank cod to the NMFS lab at Sandy Hook, New Jersey, for analysis that would, if things went well, identify distinct elemental profiles for those two stocks, and perhaps also for eastern versus western Georges stocks. Then, in the spring, he would go out with some of the Gulf fishermen he'd gotten to know, meet the school cod that the fishermen said came in from Wildcat Knoll, and see how their otoliths compared to those from the Georges and Gulf fish. "If they're a distinct stock," he said, "the otoliths should show it."

Was he afraid, I asked him, that he would run into some controversy or resistance within the Science Center over this stock issue?

"I don't know," he said. "There's a lot of bickering over stock issues.

---

*Each of the four main techniques had pros and cons that could either illuminate or tarnish findings in a given project. The best technique for stock delineation in cod seemed to be one called inductively coupled plasma-mass spectrometry (ICP-MS), in which a quartz torch, argon gas plasma, and temperatures higher than 5,000 degrees Kelvin (about 8,500 degrees Fahrenheit, or 4,700 Celsius) were used to produce distinct, intense spectra of light associated with the various marker elements. The other techniques are similarly esoteric.

And the cod team has not been real hot to do this. The whole thing could easily unravel. But at this point I've got the go-ahead from Mike Sissenwine [the center's director], the fishermen seem willing, and the guy down at the lab is ready to test the otoliths I'll send him. If nothing else it'll give us a chance to work with fishermen, maybe generate a *little* goodwill. What happens after the results come in is anybody's guess. My hunch is that when everything shakes out we'll be pretty much forced to do a major reexamination of our stock picture. I'd like to see that as something we do with the fishermen. We really have not done a good job of listening to them. We've always just blown them off by saying we couldn't incorporate their anecdotal information. That only goes so far. And here's a legitimate question we could work on together.

"But I don't know whether that'll happen. The assessment people are already saying they have to be the ones to design the study if we do one. I don't think it'll work that way. The only way to do it is as a real partnership. Otherwise, the fishermen won't have any faith in it.

"So . . . we'll see. I don't have to do this. I don't need it. But I'd sure like to see it happen from a scientific point of view. If it gets too hot, I'll hand it back to Mike and say 'No thanks.'"

That was on the next-to-last day of the fall 1998 Gulf cruise. We had finished the Gulf of Maine stations a couple of days before, catching 194 pounds of cod in forty-one tows throughout Mass and Ipswich Bays and the Bigelow Bight, with not a single tow netting more than 50 pounds. Then we'd headed south, found 600 pounds in a couple of dozen stations in the closed area in the Great South Channel and on the western end of Georges, and headed back to Woods Hole. I stepped off the *Albatross* thinking things did not look good for Dave Goethel and the other Gulf cod fishermen.

3

I arrived at the winter meeting of the New England Fishery Management Council, held at the Sheraton in Portsmouth, New Hampshire, on January 28, 1999, later than I had hoped, having driven through a snowstorm from

Vermont—cars spun off the edges of the Interstate every few miles, a split-ting vortex of snow in front of me—and so had to find a space at the far-thest edge of the parking lot, which was packed with pickups. There were usually a lot of pickups and big utility vehicles in almost any parking lot at the time, but it was clear that most of these belonged to fishermen, for buried beneath the snow in the beds were gillnets and upside-down fish boxes, and the rear windows and bumpers of quite a few sported a recently popular *National Fisherman* bumper sticker that read "Who says there's no fish?" I parked, walked a quarter-mile through 6 inches of snow to the hotel, banged my cold-creaking dress shoes clear of snow, and went in and found my way to the conference room.

Immediately inside the door, which opened to the rear and at one side of the big room, stood about fifty fishermen, most of them from Glouces-ter. In their midst, I found Joe Sinagra, a smart, funny, and eminently like-able fisherman I'd come to know. Joe found humor in even the grimmest and most outrageous ups and downs and injustices of his fishing life. His stories of his adventures trying to create alternative fisheries—markets for sea creatures people weren't yet eating—were some of his best. He had once built up, at his expense and substantial risk and in the face of consid-erable ridicule from his dockmates, a hagfish fishery in which he caught the slimy creatures in barrel traps and sold them to Asian markets desiring the hag's supposedly aphrodisiac powers. "Don't ask me," he said. "They're from the sea, and they're strange, and if you look at them the right way, I guess they're kind of phallic. It went well for a while. Then other people caught on and the market got crowded, and then you could only sell the *big* ones. It got to be a size thing." He would later almost go broke trying to develop a fishery in spider crabs that didn't work out.

Joe once entertained me with the story of how he tried to get into whelks, which are shellfish similar to conches. He started fishing for them under a Massachusetts Department of Marine Fisheries experimental fish-eries permit, which he got only after proving whelks were not conches. (He proved this by taking a whelk to a Department of Marine Fisheries biolo-gist. The biologist opened a text on mollusks, looked at the entries for conch and whelk, and confirmed Sinagra's specimen a whelk, *Buccinum undatum*. "Then it was a whelk," said Sinagra.) The department thereafter knew Sinagra as the guy who fished for whelks.

One day, some Korean business people interested in developing a whelk fishery inquired with the department about whelks, and the department sent them to Sinagra. "Probably, you know, figuring they're doing me a favor," Sinagra said.

"So these Koreans come along. The guy says, 'We want to process twenty-five thousand pounds of whelks a day.' I said, 'I don't think there's that much out there. Not to last any time.' He says, 'Oh, we're not going to just do it inshore. We want to do it on Georges, too.' I said, 'How you going to do that? Those are federal waters. You have to have an American boat there.' He said that they understood that if there's a stock of fish we're not utilizing, a foreigner can apply for a special permit. So they were looking to do that.

"So now he wanted to know where NMFS was. I took him up there, to the Gloucester office."

When Sinagra thinks something is particularly ridiculous or outrageous, his voice sometimes jumps to a higher register for a word or two, maybe just a syllable. That started to happen at this point in the story.

"So these Korean guys start talking to the NMFS guys, telling them about their great plans, and the NMFS guys are nodding their heads. Like, 'Listen to this—this is really something.' So the NMFS guy says to them, 'This sounds very interesting. This sounds great! Maybe we could help you develop this thing. Hey—we could give you some money!' And I'm *looking* at him and I'm trying to *kick* the guy under the table and say, '*What* the hell are you *doin*'? You're gonna give these guys *money*? Why can't *I* ever get any money?' Oh, well, sorry, Joe, you have to go apply for a special grant. But these guys walk in from a whole different country and they're ready to roll out the red carpet. They're ready to cut 'em a *check!* It's like, Where do you *live?*"

He threw up his hands and laughed.

He lacked such humor when I met him at the council meeting. He looked at me and gave me the sort of nod and grimace-smile that you give someone at a funeral, acknowledging an acquaintanceship amid dark circumstances.

"How's it going?" I murmured in an appropriate hush.

"Bad. We're fighting for our lives here, Dave," he said. "Fighting for our fucking lives."

The statement struck me immediately as both hyperbole and the genuine sentiment of a good many of the people in the room. The fifty or so fishermen Sinagra was with were standing even though thirty to fifty of the room's two or three hundred chairs were empty. The fishermen stood, I concluded, partly because they felt slightly out of place (though perhaps everyone feels slightly out of place in such places), a few of them wearing ties, which probably made them feel particularly at sea, so to speak, while those in jeans and flannel shirts or sweatshirts probably were more comfortable in their clothes but equally uncomfortable in the room. One fisherman, perhaps trying to split the difference, wore work boots and jeans and a gray sweatshirt and a long, elegant camel-colored overcoat. Somehow this came off successfully here, though it would have looked ridiculous on a boat.

They stood, too, I realized as the day went on, because they were too angry and anxious to sit and because standing made their anger and anxiety harder to ignore. They weren't going to take this sitting down. Along with the fifty or so at the rear flank of the room near the door, a group that fluctuated between about ten and twenty stood toward the rear of the center aisle—a location suggesting they reserved as one "public comment" option a headlong charge up the aisle, over the comment table, and into the large quad framed by the big U-shaped arrangement of tables at which the council members sat. The council members, not surprisingly, looked under siege. Strong lights shone on the comment table and the council members, and at any time there were two or three TV news cameras running, taking establishing shots of the council, the fishermen, the packed room, the state troopers leaning on the back wall, and the gray-suited private security people leaning against the side wall between the council at the front of the room and the fishermen at the rear of the room. Though close to the size of a hockey rink, the room was overstuffed and overheated. We sweated in our clothes as outside it snowed.

The council was, as one member put it in a position paper distributed in the lobby, "way beyond the rock and the hard place." In the previous weeks, the council's Multispecies Monitoring Committee (composed largely of staff from NMFS, the council, and state fisheries agencies), having digested all the Northeast Fisheries Science Center's latest information regarding groundfish, including the dismal catches my boatmates and I had collected in the Gulf cruise some eight weeks before, had concluded that several key groundfish species were in deep trouble and seriously overfished (not new) and (new) that Gulf of Maine cod were at a record low after a year when even the restrained fishery had taken twice what the stock could stand. Under the Sustainable Fisheries Act (SFA), the council could not allow the Gulf of Maine cod population to drop below 7,500 metric tons, and it was down to around 8,300. The cod harvest therefore absolutely positively had to be cut 80 percent, to an absolute maximum of around 800 tons. Because close to that much cod was taken every year as unintended by-catch in flounder and other fisheries, the council needed to completely eliminate the directed fishery for cod; it needed to ensure that no boats would go out trying to catch cod. Otherwise, the Multispecies Monitoring Committee advised, Gulf of Maine cod might collapse as drastically as had Georges cod a decade before. Also, the secretary of commerce might step in and take over management of the fishery, which if you were on the council was a humiliating prospect.

The council's task, then, was to create regulations for the coming groundfish season, which would start within a month or so as the fish began to arrive, that would entirely stop people from fishing for cod and that would also prevent people from accidentally catching cod while fishing for other fish; and the council had to accomplish this while taking into consideration, as the Sustainable Fisheries Act also required, all resulting socioeconomic consequences. Finally, because they had to do this in a duly warned public meeting and the fishing season started soon, they basically had to do all this today. While seated before, and taking public comments from, this crowd. Piece of cake.

Before the council were three proposals aimed at meeting the goals established by the Multispecies Monitoring Committee. One, produced by the council's plan development team, recommended expanding the regulatory regime currently in use. It would use limits in days at sea and daily catch combined with a set of expanded rolling closures that would create a wide, groundfishing-free zone that started in Massachusetts Bay and slid north along the coast of the Gulf as the season progressed, like a fat ruler sliding up a map, to provide the migrating fish a moving refuge. Coming from the council staff, this plan did not attract any passionate advocates, and it worried the small-boat fishermen like Dave Goethel, who would be most hindered by the expanded closures of inshore waters.

The other two plans were bolder and more sweeping. One, from the Maine Department of Marine Resources, called for banning groundfishing all spring and summer in all water within 40 miles of the coast between Cape Cod Bay on the south end and around Portland on the north end—in other words, all inshore Gulf waters south of Portland, including all of Massachusetts and Ipswich Bays. This would greatly relieve pressure on the cod. But it would also force all dayboats between Cape Cod and Portland either to stay home or fish at hazardous (not to say exquisitely uncomfortable) ranges—some five to six hours from shore and in dangerously open water. It would essentially end the Gulf's small-boat groundfishery. For obvious reasons, most of the small-boat fishermen between Cape Cod and Portland hated this plan.

The third proposal, perhaps the most original, came from the Gulf of Maine Fishermen's Alliance, or GOMFA. This two-year-old group had put together a fragile but for now cohesive alliance of gillnetters, trawlers, and longliners who were trying to set aside their long-running turf battles and squabbles regarding gear type to present a Gulfwide, united-we-stand approach. Most GOMFA members fished inshore, and they had offered a proposal that for once addressed the reality that small dayboats could apply fishing pressure only throughout their very limited daily ranges, whereas most large boats could apply pressure (and make a living) throughout the Gulf, not to mention Georges. The GOMFA plan called for dividing the Gulf into inshore and offshore fisheries and requiring every boat to limit itself to one or the other for each year's groundfishing season. Any boat could choose either inshore or offshore, but it had to fish only those waters (for groundfish, anyway) all year. Since most of the large boats would want to reserve the option to fish offshore, this plan would, GOMFA predicted, reduce the fishing pressure inshore by preventing the concentration of all boats on the cod schools as they moved along the coast. Daily take limits could therefore be higher (and could be adjusted as necessary as the season went on) and still meet the overall catch reduction requirements. The plan had the strength of simplicity and fairness—everybody got a choice of turf and the right to fish. It carried two critical weaknesses: It was difficult to predict the exact effects the plan would have on effort and by-catch; and the long-range boat owners (including at least one powerful council member) would probably hate it.

These proposals presented the fishermen in the room with a Scylla-and-Charybdis dilemma, with a third ugly monster thrown in. The resultant tension showed, especially in much of the contingent from Gloucester. The biggest fishing town in the Gulf, Gloucester is a closely knit town that had long represented itself ardently and effectively at council meetings. So it did here. Most of the Gloucester fishermen hated the DMR inshore-closure plan, for it closed the water right off Gloucester, which all of them fished often, and would leave the small-boat owners among them nowhere at all to hunt groundfish. As a group, however, they seemed of mixed mind about the GOMFA inshore-or-offshore option, for many of them had long-range boats and didn't like the prospect of having to decide between fishing the moderately productive and relatively uncrowded banks in the Gulf's eastern half and the often richer and more convenient, if more crowded, waters in Mass and Ipswich bays. They expressed similar reservations about the council staff's rolling-closure plan. Some of them, grumbling at the back of the room, advocated a sort of scorched-earth retreat, saying that if there were so few damned cod and the council was going to tell people they could not fish certain places, why not just shut the whole Gulf to all groundfishing, or hell, all fishing while you're at it, and give the place a rest? You did it on Georges, why not here? "Close the whole fucking thing!" as one yelled. "But better pay us," grumbled another. "You're going to lay us off, you have to pay us. That's how it works."

It all struck me as a rather unpromising way to create conservation policy. Balancing the socioeconomic and ecological needs appeared impossible; their combined weight was too much for the fulcrum. And the room stank of dysfunction. No one trusted anyone. All the council members looked exhausted or scared or depressed or angry or all three, and most of the audience appeared on the verge of rage or despair. Given this, the protocol—the taking of public comment on the proposed plans, followed by open debate among the council, followed by council motions, discussions thereof, and then votes—proceeded haltingly. When I got there, the council was still taking the initial public comment. A few fishermen spoke in favor of one plan or another (a couple of them spoke quite eloquently in favor of the GOMFA plan), but most complained that any of the plans would bring the fishing communities great suffering. Dissenters from this line tended to get razzed, sometimes obscenely. A ponytailed fisherman went forward, sat at the mike, and said he recognized everyone would need to cut back on the cod and that he was willing to do so, and that it was time for fishermen to take the hits necessary to let the fish recover. As he got up, someone yelled, "Pull his ponytail!"

Some of the fishermen did soberly address the assessments and proposals. One, referring to the DMR plan to close all inshore waters, said, "The message you'll send if you adopt this plan is crystal clear: Bigger is better. This proposal's closure discriminates against smaller boats while leaving offshore boats plenty of options to go elsewhere. This adds to the impression of a conspiracy to eliminate the inshore fleet." Others, including some on the council, expressed doubts that the GOMFA inshore–offshore plan would prevent a directed fishery or otherwise control the cod take. One fisherman—the only one I heard who took up scientific issues in a way other than attacking NMFS's twice-yearly surveys—said he worried the inshore restrictions might be overdone because he believed NMFS's two-stock theory might be underestimating the number of inshore cod: "All these plans are flawed because they assume we have just two stocks in New England. But to classify the New England groundfish as just two stocks is absurd."

By one o'clock, all the plans had taken such a beating that it seemed they could hardly stand. The room was tired. Hungry, I was thinking that this was no business to conduct on an empty stomach. People were irritable enough as it was. Dave Goethel, up in the front row, looked tired and slightly ill (about the way, I realized, I must have looked on his boat). The Gloucester fishermen looked angry and implacable. The council looked more beleaguered and discouraged than ever, as well they might: With the day half over, the only thing they had accomplished was to leave both their main new options in tatters. If neither of them passed, their only remaining option would be the expanded version of the present regulations, which no one liked and which the year before had allowed twice the allowed cod harvest. It was difficult to imagine what they could come up with that wouldn't risk either a riot by the crowd or an overrule by the secretary of commerce.

Finally, toward two, executive director Paul Howard, realizing they weren't going to get in a vote before lunch, gaveled in a lunch recess.

The fishermen dispersed, many of the locals heading out the door. Others, particularly those who had spoken most passionately during the morning, were buttonholed in the hall by reporters looking for interviews. Dave Goethel, who had spoken before I arrived, was talking to a pair of reporters, saying that while he saw the need to reduce fishing, he hadn't seen the decline in cod in his home water that NMFS had described. When he started talking about the two-stock theory, one of the reporters interrupted him and asked him what the closures in the various plans would mean for him. A few feet away, three reporters were gathered around one of the Gloucester fishermen who had pleaded with the council not to be put out of business. His buddies next to him were amending some of his statements and talking about being crushed and ruined. The reporters scribbled hurriedly, their eyes bright, on the trail now of a familiar story line they knew would sail. Those who had to file that afternoon were realizing their deadlines would come before the council made a decision, and they were forming their leads: Fishermen plead for lives as council dickers.

"Seven hundred boats," I heard one fisherman saying. "There's seven hundred boats that work this water they're talking about closing. We're looking at an entire way of life being squashed by these people."

Beyond the windows, the snow still fell. I found a phone and made some calls, then searched for the cheapest place in the hotel to eat lunch. Twelve bucks got me into the bar and access to a buffet. It also got me Dave Goethel, who was at the bar drinking a beer and talking quietly to a couple of fellow fishermen on either side of him. Goethel was wearing gray slacks and a navy blue blazer and a dress shirt and a tie that was a fish. "I don't know what kind of fish this is supposed to be," he said. "A sucker fish, maybe. We're all such suckers." The two fishermen he was talking to laughed. I had not talked to Goethel in several weeks, and he seemed a bit distant, as if perhaps he saw me as part of the problem or feared his compatriots there would. I didn't blame him; most of the press had covered the groundfishery's collapse poorly. Few writers bothered to paint a picture involving more detail or depth than cold-hearted scientists and bureaucrats slowly choking virtuous fishing families to death or rapacious fishermen sacking the sea. They rarely conveyed the sense that either side ever acted on poor vision or bad judgment or unwittingly sabotaged its own interests, and they seemed blind to the cultural, scientific, and policy conflicts that made the situation so difficult to resolve. The general level of reporting had moved maybe a half-inch up the scale of sophistication and insight from spotted-owls-versus-loggers. This was, unfortunately, par for the course for reporting on resource-extractive issues.[*]

Leaving my notebook in my pocket, I asked Goethel how his comments had gone that morning. He said okay; he'd given them his two cents but didn't expect it to do much good.

"Maybe speaking early wasn't the luckiest timing on this particular day," I said. "A lot of noise in the meantime."

"No, I don't expect anything really good to come out of this thing," said Goethel. "But I'm tired of having the burden of proof on the fishermen. I'm willing to take whatever cut in daily limit they decide we need. Err on the side of the fish this time. Just don't force me to travel clear across the Gulf to fish. I can't hack that like the big boats do. Too many people operating on too thin a margin of safety already. Somebody's gonna get hurt or killed if we have to sail six hours to fish."

I asked him what he thought of the GOMFA inshore–offshore plan.

"That'd make things easier for me," he said. "And it's remarkable they got so many different gear types to agree on the thing. First time *that's* happened. But I don't see the council passing it. The big boat owners on the council won't let that happen.

---

[*]A few notable exceptions broke this rule of facile mainstream-press reporting on the groundfish collapse, most notably the work of the *Boston Globe*'s Scott Allen, the *Maine Times*'s Phyllis Austin, and freelancer Susan Pollack. Penetrating, constructive contributions also came from a few writers, such as Paul Molyneaux and lobsterman/reporter Peter Prybot, in their work for such small-industry publications as *Fishermen's Voice* and *Commercial Fisheries News*. As noted earlier, the *National Fisherman*, the leading industry journal, usually worked its New England fishery beat thuddingly, apparently more intent on lambasting NMFS's science and the council's regulations (and rarely noting the difference between the two) than in promoting the deeper consideration of issues that was so obviously needed after more than a decade of gridlock. The press bungled this story at least as badly as did government and industry.

"But hey," he said, lifting his beer and raising it in mock toast to his two companions. "Maybe they'll surprise me." His companions guffawed without mirth.

I left Goethel and his friends, served myself some lunch, and sat down at a table. A wall of windows on one side of the bar looked onto a small courtyard into which snow spun. Though calmer than the conference room or halls, the room held an uneasy energy, with complaining tones rising among the voices and a couple of people, maybe having had one too many, talking louder and laughing derisively at jokes I couldn't hear. The light was foul: not stark as in the conference room or overly cool and white as in the lobby, but a depressive dim gray. It was the light of a dusty closet. The snow outside held a warm luminance by comparison. I thought of what it would look like disappearing silently into the sea's dark swells.

Back in the conference room, the lunch break, which earlier had seemed so necessary, appeared to have been a mistake—as perhaps was appropriate for a ceremony presiding over a crisis in which so many desperate acts had gone wrong. The murmur from the Gloucester fishermen by the door had grown to a low rumble. The council members, taking their chairs, seemed more tentative and reluctant than ever. As people took seats, the group of fishermen at the back of the center aisle grew to its largest size yet, and some of them at the rear, in an effort to see better over their friends, pulled chairs into the aisle and stood on them. These burly fishermen all pressed together, a couple of those in back resting their hands on the shoulders of their fishing brethren before them, made a particularly picturesque tableau with a sort of Minuteman flavor to it; they were thus arranged for about a minute before one photographer and then several recognized their day's best photo op and crouched in front of them to fire off a cluster of shots. The longer the cameras flashed, the more defiant and stoic the fishermen looked. The police and guards took up positions along the side walls.

The council, having heard enough before lunch, tried to buckle down to business. Eric Smith, who as the assistant director of the Connecticut Marine Fisheries Office could at least claim relative disinterest in Gulf of Maine affairs, made an argument for and then moved for adoption of the

GOMFA inshore–offshore plan. With that motion seconded, the council rehashed the arguments for and against the proposal. An hour later, after a few amendments had been added to address concerns about whether the plan would encourage a directed cod fishery, the council voted. As they went around the table it became clear it would be close, and it was: It failed, seven for and nine against, as Eric Smith, apparently having changed his mind during the debate, joined the nays.

When the tally was announced, some of the fishermen started yelling. "Five days we took off for this crap!" "We're tired of this bullshit." "I didn't come up here to get sodomized!" One of the fishermen in the group in the aisle stepped forward, shouting and gesturing, but retreated when one of the security guards moved his way and asked him to step back please.

For the moment, the grumbling stasis held, allowing the meeting to continue. Someone moved for adoption of the DMR plan and was seconded; after a much briefer discussion, this plan also failed, seven to nine.

The council now had to decide between taking up its own planning unit's proposal, that of the expanded rolling closures and catch limits, or putting together something different. Perhaps because the rolling-closure proposal had already taken a beating that morning, they chose to ignore it for the time being and try to improvise instead. For the next three hours, as the fishers looked on and grumbled and dinnertime neared and the snow outside fell unseen and somewhere in the dark the cod began to move toward Wildcat Knoll, the council tried to hammer together a plan from pieces of the maddening, maddeningly ineffective existing regulations; the least offensive and least effective parts of the two defeated plans; and a hodgepodge of technical tweaks and exemptions. Witnessing this was like watching a group of strangers attempt to hastily arrange a meal for an important state dinner after half the guests had rejected the menu. There was no central component or overriding principle—no entrée around which to organize everything, much less any agreement on what kind of cuisine to draw from. Someone would plate an idea, and if it made it around the table without causing anyone to gag, it was added to the menu.

By six o'clock, they had this compromise proposal, as they were now optimistically calling it, ready for public comment. As the plan's restrictions were mild compared to those put forth earlier, some of the fishermen went to the mike to express support. The most entertaining plea came from a fisherman who looked to be in his late twenties or early thirties and who pushed aside the chair in front of the table, dropped to his knees, and began, "I did it for my wife; I'll do it for you. I'll get on my knees and beg." This comic relief was soon spoiled, however, when ugly jeers greeted the comments of representatives of two environmental organizations who went to the mike to warn that this plan could not possibly meet the harvest-reduction goals the council had earlier acknowledged as necessary. When the second of the representatives, a woman, finished and rose to return to her seat, someone yelled, "Go wash dishes!"

The debate moved to the council. To the obvious surprise of many audience and some council members, the central point of debate became whether it was time, finally, to err on the side of the fish, the consequences be damned, so that for once the onus would not be on the fishing industry—much the line of thought Goethel had expressed in the bar. This obviously meant defeating the compromise proposal. Council member John Williamson, a sometime fisherman who clearly recognized he would infuriate some fellow fishermen by doing so, made the most passionate plea for erring toward conservation. "We're here because for ten years the council has always accommodated those who claimed hardship. It's the same every year, and we've always caved, and we haven't helped a thing. We've made things worse. We have reached the end of the line. We have no more room. We cannot approve this plan."

"Traitor!" someone yelled. Williamson looked at the crowd with resigned dismay. A half-hour later, the council defeated the plan with the third nine-to-seven vote of the day. One of the Gloucester fishermen, a man I'll call Tom, who was usually diplomatic and reasonable, walked up the aisle shouting at the council and, breaking protocol, stepped around the table where the speakers from the audience had sat to use the microphone. With that, he entered the big square space framed by the council's horseshoe arrangement of tables. One of the security officers also entered the little quad, and, taking the kid-glove approach, crossed to Tom, spoke very quietly to him and—his one mistake—put his hand, gently, on Tom's arm.

The moment he touched him, the dozen or so fishermen at the rear of the aisle surged up the aisle toward the council. The three security officers who had been at the edge of the room all day immediately moved to intercept them. "Stop right now," the one closest to the front of the room said. He was walking at an relaxed but brisk pace along a row of seats toward the aisle, and without shouting he somehow spoke loudly and forcefully but with an edge of sympathy and brotherly appeal. He looked remarkably at ease and confident considering he was one of three people approaching more than a dozen large, angry men who had three or four dozen large, angry friends forty feet away. He spoke loudly but moved as if nothing was terribly wrong. Most of the fishermen hesitated, but one continued up the aisle. A couple of people sitting on the aisle got up, hurriedly grabbed their most essential belongings, and moved away from the aisle. I was a little closer to the potential action than I wanted to be myself. "Please!" the security man shouted, and he stopped and held up one hand in a gesture that was both an appeal and a stop sign. If he'd yelled "Hey!" or continued forward, we'd probably have had a riot. Instead, looking a bit surprised, the fisherman stopped. The room was dead silent and dead still. At the front of the room, the security man who had intercepted Tom had taken a step back from him, and Tom was holding his hands up, too, as if to say it was all right, he was not trying to make trouble.

The security man spoke again, more quietly. "We don't want to go any further here. Let's just all back up." And everyone did.

They finished at one the next morning, when the council, having regrouped after a late dinner and taken up the rolling-closure plan, passed a "framework adjustment" (an addendum to existing regulations) that expanded the previous year's rolling closures and cut the daily catch limits for Gulf of Maine cod to 200 pounds. Because some council members doubted the measures would adequately reduce the catch, the regulations authorized the NMFS regional administrator to cut daily cod limits to as low as 5 pounds (about one fish) if landings exceeded 402 tons, or 51 percent of the season's total allowable catch. It wasn't an ideal plan, and it was going to hurt dayboats more than overnight boats, for the closed areas, at about 30 to 60 miles north to south and almost 100 miles wide, were big enough to encompass most or all the water a dayboat could comfortably cover. Still, the plan seemed to have a good chance of meeting the biological goals. It may not have been fair to all the fishermen. But for once, the council could credibly argue it had erred on the side of the fish.

PART V

I

Dave Goethel, sitting at his kitchen table with a copy of the new regulations a few days later, could see right off that April and May, when the rolling closures would forbid him from groundfishing in his home waters, would pinch. Like most Ipswich Bay fishermen, he usually spent those months fishing for incoming cod, flounder, and other groundfish, sailing south an hour or two to meet the fish in April, then fishing them ever closer until he was directly off Hampton Harbor as they moved past in May. He typically earned 25 to 40 percent of his year's income in that six to eight weeks. With the closures, he couldn't. This worried him. Cuts in fishing days and other restrictions had already halved his income over the previous three or four years. He had yet to finish a year in the red, but this would take him close. What was he going to do in April and May? He could shrimp through March, as was his custom lately, but then what? One possibility was to steam east two hours each day to Jeffrey's Basin, a good groundfishing area that would be just north of the closed area in April and which usually had some flounder moving in by then. When the closed area jumped north 30 minutes of latitude on May 1, covering those grounds, he'd have to go four hours south to fish legally for groundfish, below 42 degrees, 30 minutes (or forty-two thirty, to use the vernacular). So that was the plan: Shrimp in March, hunt flounder off Jeffrey's in April, and in May motor four hours south to Mass Bay for more flounder.

He would go without Charlie, for after learning of the new regulations, Charlie had cast around for work and taken a job at a car-detailing outfit. "Probably the smart thing for him," Goethel said later. "With the catch limits, he'd be lucky to pull a hundred a day, and that's a twelve-,

165

sixteen-hour day. He'll make more per hour at the detailing place. Was too bad. He was a good guy and good crew."[*]

So Goethel fished alone. Things went okay for a while, even though catching shrimp, he says, is like catching clouds, for they form, disperse, disappear, and regroup in manners even more strange and unpredictable than those of groundfish. It had taken him several seasons to get the feel of where to find them. He was having good luck this year, though, and prices were holding, and as he usually fished alone and did not have crew to pay, he managed to get through February and March in the black.

By the last week of March, however, he had used up all his shrimp days, so he rigged his flounder net and began steaming each morning to Jeffrey's Basin. He pulled a few good hauls, but most days he was catching only a few hundred pounds of flounder and nowhere near his limit of 200 pounds of cod, which was the fish bringing the most money—the new restrictions had pushed prices up nicely, to well over a dollar a pound and sometimes close to two for the biggest cod. But the fish didn't seem to be there yet. After a couple of weeks, taking into account fuel costs of more than a hundred dollars a trip and even the most modest nod toward wear and tear on the boat, Goethel saw he was not making anything. He began to wonder if he should start sailing all the way down to Cape Cod Bay to fish for flounder below the closed area. It was a six-hour trip each way, but some of his friends in Gloucester were doing well down there, landing several hundred pounds of flounder every day and one or two hundred pounds of cod and returning home with a load worth a thousand dollars or more. Now and then they were landing loads of dogfish running several thousand pounds, real net-busters—unpleasant, but at that volume a good payday.

All this hardly seemed magnificent compared to years past, but a series of four-figure checks at this point sounded good. Dave talked it over with Ellen and decided to give it a try. He'd steam down to Mass Bay and fish

---

[*]On a small dayboat like Goethel's, a typical one-person crew usually is paid 20 percent of the day's gross after fuel costs. With the various restrictions and the main money-fish, the cod, essentially off-limits, a normal day's gross in 1999 came to roughly $500. Subtract about $75 for fuel, and you end up with $425, a fifth of which is $85.

for three or four days at a time, sell his fish every evening in Gloucester Harbor, and sleep there on the boat. After three or four days, he'd return home for a day or two, then do it again. Not pretty, but he could stand it for two or three weeks.

He bought a permit to sell his fish in Massachusetts, stuffed a bag with changes of clothes and a toothbrush (he'd find a shower at the marina somewhere), filled a box with coffee and stovetop meals for his hotplate, and set off. The plan worked, sort of. Nearly every day he caught several hundred pounds of flounder, and he usually got his couple hundred pounds of cod as well, sometimes even had to throw a few over. A few times he hit big packs of dogs and, beauty be damned, dumped them into his fish boxes and took them to market.

The sleep-at-the-dock plan, however, failed miserably. "You can't believe how much noise is in a harbor," he said later, "till you try to sleep there." Every twenty minutes some racket broke out. A returning boat chugged in; somebody's boat dog barked; a truck pulled next to the docks and honked to pick someone up; a generator engine fired up; somebody working on a motor banged tools around; and beginning around three or four in the morning, crews started clomping down the piers and tossing their bags and supplies onto boat decks and starting engines and shouting instructions to each other. "In three days," said Goethel, "I was a zombie."

After two or three such trips he was glad to see May arrive. The southern edge of the closed area jumped a half-degree north, bringing it within a four-hour steam, a long but feasible round trip for a dayboat. He would rise at two in the morning (two hours before his usual time), cast off around two-thirty, motor south to clear the closed area, and set his first tow around dawn. Then he'd try to fish three tows, which would take him until two or three in the afternoon, and head home. After selling the fish around dinnertime, he'd get home about nine, eat, and go to bed. If he had sold no more than 200 pounds of cod, he might get up four hours later and do it all again. If he hit a lot of cod, he'd keep 400 pounds instead of throwing 200 over half-dead and take a penalty day the next day to stay home and sleep. He was allotted eighty-eight days of groundfishing (the limit established by the New England Fishery Management Council since 1994), so taking the penalty meant losing an entire day's worth of flounder income that he couldn't replace. But at least he got some sleep and avoided the aggravation of tossing cod overboard. He hated discarding cod just to stay under a daily limit, for it hardly saved the discarded fish, over half to three-quarters of which would die from the stress of being caught, compressed in the net, and pulled up through several dozen fathoms and dumped in the boat. He knew the limit was intended to discourage the targeting of the fish in the first place, but when you were catching fish you couldn't avoid, throwing them overboard simply wasted them. He hated it. He knew fishermen who were not bothered by it, and he knew some were intentionally focusing on cod to capitalize on the good prices, going out once a week and catching one to two thousand pounds and all the flounder they could haul and in exchange taking a day's penalty for every 200 pounds of cod past the first 200. He wouldn't do that any more than he would fish at night. It defeated the purpose of all the sacrifices everybody was being forced to make.

What made it all the more aggravating was that the cod were difficult to avoid that May. He had to work to find spots where he could make three tows of flounder without hitting his cod limit. They were running thick.

Back home, Ellen worried more about Dave than she had in a long time. Going out so far into unfamiliar water, tired most of the time and pushed to make every day pay, he was squeezing thin his margin of safety. Fishing alone was bad enough: No one else on the boat to watch his back for freak waves or swinging net gear, to help judge the weather, to throw a ring and turn the boat around if he fell overboard. The distance hurt, too, because he could not see and fully judge the seas until dawn (the moon was deep into its wane at the time), and by then he was far from shelter. Though he had sailed the boat for years, often in the dark of early morning, it was still difficult to tell an 8-foot wave from a 12-foot wave by feel alone. The difference was critical, for a boat can generally be capsized if caught broadside by a wave taller than the boat is wide. The *Ellen Diane*, with a beam just under 15 feet, therefore ran exponentially more risk as seas neared 15 feet. Even smaller seas could swamp a boat that has lost power or suffered damage. Goethel feared reaching the open area and finding the seas heavier than expected, 10 to 12 feet instead of 6 to 8, nearing his zone of discomfort, or getting two or three hours from cover and having a big storm brew quickly. The fishing press (and after the popularity of *The Perfect Storm*, much of the coastal mainstream press) had recently run several stories about fishermen who, forced by tight margins to fish alone and pushed out of familiar waters by fish shortages or regulations (fishermen like Goethel), had sailed too far from home in iffy weather and gotten into deep trouble.

On the upside, he was catching fish: healthy yellowtail flounder, many of good size, inevitably complemented by some nice cod and now and then a net bulging with dogs.

"What passed for good fishing," as he put it.

"Voracious almost beyond belief," as Bigelow noted, dogfish will eat just about anything, and when schools of baitfish are plentiful they pursue them (baying like hounds, one supposes) in large packs. In spring and early summer in New England, they're often chasing two of New England's most common baitfish, herring, or sand lance. The herring, with its familiar, archetypal fish shape (picture the Christian fish symbol and you've pictured a herring), runs between 12 and 16 inches long when grown. The sand lance, 4 to 6 inches long, looks more like a tiny eel, which it also resembles in its undulant swimming motion; thus its alternate moniker, *sand eel*, used by most fishermen. *Sand* refers to its preference for sandy bottoms, into which it can burrow with its sharp nose, seeking cover, and disappear in less than a second.

The sand eel is the less pleasant meal (not as much meat, and a nose so sharp it sometimes punches through the stomach lining of the fish that eat it), but both it and the herring are so widespread and numerous in New England waters that no predator can afford to pass them up. Striped bass and silver hake will feed in a frenzy on either species, chasing them into shallow water and even up onto beaches, where both predators and prey may strand themselves on the shoaling bottom. "We saw this happen once at Cohasset in Massachusetts Bay," writes Bigelow in *Fishes*, "when hake and herring were so intermingled in shallow water at the height of the carnage that we soon filled our dory with the two, with our bare hands."[1] Dogs and cod generally hunt these species less visibly, gobbling their way through schools near the bottom or in midwater.

If you're looking for dogs (or cod or hake, for that matter), you're

often looking for herring or sand lance. So when Goethel had trouble finding flounders that May, he would sometimes watch his fish-finder for
clumps of color with the particular profile, striations, and density—a thin-
ner line, a finer warp than cod or herring—that suggested sand lance. If he
could find the lance, he figured, he might be near the dogs.

One day, about a week into May, Goethel set out for Mass Bay at 2:30 A.M.
and, hitting the open water outside Hampton Harbor, encountered seas
running 6 to 9 feet. Or at least that's what they felt like; it was a moonless
night, utterly black, and he could not see past the bow. He wasn't overly
concerned, as the weather service had called for waves that size, but he
found it slightly troubling that the waves were that large so early. He
didn't like to work in double-digit water.

As he sailed south the waves seemed to grow, though he wasn't sure. A
few broke over the bow and washed over the deck. Yet that had happened
plenty of times before, and the boat still felt good. Maybe the waves had
just shifted.

About an hour later, around four, he realized he was seeing far fewer
other boats than he usually did by then: no lights off to either side or
ahead. Nothing on the radar. Apparently all the other fishermen had
decided it was too rough to be out. And it was getting rougher. The waves
were smacking him harder, some of them breaking over the bow or, more
disconcerting, lifting the boat and then delivering a whack from the side.
They slammed into the cabin with boat-shaking booms. When he turned
on the spotlight, he saw waves that easily blocked the horizon, waves you
had to look up to. When he'd top the wave and slide down the other side,
the light would show first spray above and then the swirling trough, and
he'd slide down into a deep trench of water. It looked bad. It was big and
jumbled and steep, the sort of water that could make real trouble. But he
knew the spotlight always made things look bad, and those frame-rattling
waves aside, the *Ellen Diane* still felt solid. There was a certain rare pleasure, actually, in feeling the boat handle it so well.

Yet as he neared the fishing grounds he grew increasingly uneasy. Like
a driver passing one exit after another on an icy night, driving steadily but

feeling the tires slip now and then, he knew he was pushing his luck and that the most sensible thing to do would be to turn around. He kept going out of momentum and for the tense, testing pleasure of engaging the storm.

When dawn showed the water, he found he was almost completely alone. The only visible vessel was a distant freighter.

He could also see that the seas were not 6 to 9 feet but more like 12 to 15. Distinctly double-digit. Knowing it was the kind of decision that sends boats to the bottom, he resolved to make a tow. Just one, to pay for the fuel. He found some sandy bottom of the sort that often held flounder, eased the throttle, set the boat on autopilot, and began the routine of setting net. The seas made the work decidedly unroutine, however. Just getting the hatch open to go below and switch on the winch hydraulics was touchy, with the big hatch door heaving and slamming onto the floor when he opened it and the boat pitching as he went down the ladder. He came back up and closed the hatch, donned his foul-weather gear, and opened the cabin door to go let the net out.

The deck was frightful. It was heaving and rolling and lurching, and seawater gushed over it and poured in great riverlike torrents over the stern. Worried about being knocked down and washed overboard, Goethel crawled on all fours to get from place to place, hanging on to anything he could—the winch housing, the winch cable, the gunwale, the transom that held the net spool. Waves broke onto his back and brought the sea swirling around his hands and knees. Then came the bad part. To get the net started he had to stand up, one arm hooked around the transom support, then reach up and pull the cod end off the spool overhead. He was right by the stern, and if a big wave hit him, he'd probably be gone. He held on, though, and got the end of the net into the water. Then he crawled back to the hydraulic controls, which were sheltered behind the pilothouse, and stood to work the controls and let the net out. When the net was out, he crawled once more over the wet sluicing deck to the transom to attach the doors to the front end of the net. He hooked them to the cables without falling in, crawled fore again to the hydraulics, and finished letting the net out. Finally, he opened the cabin door and went back inside.

He secured the cabin door behind him, sank into the captain's chair, and shucked off his rain hat and jacket. He was a little amazed that he was still aboard. He wiped the saltwater off his lips and eyebrows and pushed back his wet hair.

A sudden movement out the starboard window startled him. He looked and almost jumped out of his chair. Outside the window, maybe 6 feet away from his shoulder, so close and sudden he felt it in his chest, rose a huge black form. Goethel was peering directly into the eye of a whale. The whale was rising slowly, nose skyward, its great mouth open as if it were taking a bite out of something. When it stopped rising, baleen fanned across its mouth and seawater gushed out, as a flood through a grate. Then, turning away from the boat, its long side fin swinging up out of the water like a loose sleeve, the whale disappeared into the sea. The water around it boiled.

Goethel was as astounded as he'd ever been. He looked out the front windows and saw water boiling there, too. Just ahead and to port, a huge roil of bubbles disrupted the surface. The sea broke open as another whale emerged, mouth open and its broad, wattled chin toward the boat. As Goethel passed it, the whale bobbed back down into the water.

More whales did the same—he counted six in all—and as they rose repeatedly on all sides of him, startlingly large, Goethel realized they were humpbacks that were bubble-feeding—spinning circular curtains of bubbles around a school of fish far below and then, as the confused fish gathered into a tight column inside the bubbles, swimming upward through the column with their great mouths agape.

Goethel liked whales. But today, beleaguered by the bad weather and rattled by the trip over the deck, he wished these would go away. They were almost surely swimming under the boat; what if one of them bumped him? What if one went behind the boat and got tangled in the cable? In this water (he was still climbing 12-foot waves) a good tug on the cable might pull the stern under, the rest of the boat to follow. He thought about pulling in the net but decided that would only increase the chance of a tangle; better to leave it on the bottom.

After a few minutes of feeding, the whales, apparently having got their fill, started to entertain themselves with the *Ellen Diane*. Humpbacks often approach boats closely and indulge in what appears to be play, which, along with their spectacular, full-body breaches and tail-flashing dives, make them favorites of whale-watching tours. They come directly alongside a boat, even swim beneath it, and thrill the passengers by suddenly exploding out of the water nearby and breaching with a terrific splash. Sometimes they even gently bump a vessel, giving the whale watchers a screaming treat.

Goethel wanted no such thrills. The whales did. They switched from their relatively gentle bubble-feeding rises (unsettling enough) to firing up out of the water and splashing down less than a dozen yards away. Even amid the big seas, the wakes from their impacts knocked the *Ellen Diane* sideways. Sometimes one would suddenly surface right next to him and swim alongside, lifting its head as if to peek inside, then roll away, flip its tail, and dive.

When he thought it could not get much worse, he spotted one between the cables of the net. It was following the boat. "It was like that scene from *Jaws*," he said later, "only with a whale." The whale did not munch the stern. It just swam along making Goethel nervous. After a few minutes, it sounded. Goethel watched and waited, worrying that it would hit one of the cables or, worse, swim back into the mouth of the net after the fish that were gathered in there. But the cables didn't shudder, the stern didn't pull, and the *Ellen Diane* kept towing. Around him, one at a time, the whales sounded, their backs arching with the humped-neck motion that gives them their names.

After about ten minutes, when he figured they were finally gone, Goethel crawled out to the deck, reeled in his net, and emptied a small load of flounder. After setting the net again, he sat down to sort the fish. Working on a rolling deck awash in waves no longer seemed extreme. It was a crazy day, and he just had to roll with it.

He made two more tows, both light. By one o'clock, it was clear he wasn't going to catch much; time to head back. With almost every wave coming over the bow, the *Ellen Diane* as much submarine as boat, he made for Gloucester. He arrived about six, unloaded, collected his chit, and, after hearing that the weather was to be bad for a week, turned homeward. He got to the house at close to ten.

"You won't believe my day," he told Ellen.

Three days later, the blow subsiding, Goethel headed back to Mass Bay to see if he could fish without assistance from whales. In much calmer seas and mild weather, he cast off early and steamed to the open area. On his first tow he found the latest of the surprises this strangest of spring seasons would offer him. He described it in a letter he wrote a few days later to the New England Fishery Management Council:

> On May 17th I made a tow in 40 fathoms marking [on a fish-finder] a lot of what I thought were sand eels that the whales had pushed off the bank. Because of the quantities of dogfish known to be in the area, I made a one-hour tow. Well, they were not dogfish but rather an [entire net] full of 10- to 15-pound codfish. Realizing the massive waste and destruction that was about to occur, I slit the belly of the net . . . and dumped most of the fish out the hole. I estimate that between 5,000 and 10,000 pounds swam away, and approximately 5,000 pounds sank to the bottom, dead. Out of the bottom bag [he had tied off the very end of the net so that he could keep some fish], I kept 12 boxes of cod [at about 100 pounds each] and took a six-day time penalty.

It was the biggest haul of cod Goethel had ever netted. A freak, he figured, the cod concentrated by the storm. He sailed home, put the boat in the boatyard for some needed maintenance, and took a few days off, thinking he might as well wait until June, when he could again fish close to home. Maybe by then things would return to normal.

But things continued all a-whack. For the rest of the spring, neither fishermen nor fish seemed to be where they were supposed to be. Fishermen all over the Gulf were catching cod no matter what they did to avoid them. And though Goethel, getting back on the water in the first week of June and fishing Ipswich Bay, didn't hit any more five-ton loads, he could not avoid catching several hundred pounds a day. He tried every combination of depth, topography, and bottom structure he could think of to find habitat with flounder but no cod. The best he could do was to catch them in even proportions. The day's first haul was usually the worst. He had once called it the glory tow. Now it was a nightmare, often bringing up as many as 500 pounds of cod amid an equal or lesser amount of flounder. Every fishing boat he talked to was getting similar results.

The carnage worsened on June 1, when the reported catch passed the trigger point of half the season's total allowed catch, leading the NMFS regional administrator to cut the daily cod limit for each boat from 200 to 30 pounds. In theory, this would reduce cod take and mortality. In reality, it just compelled the boats to throw away more fish—usually 170 pounds more. "It's a complete disaster," said Goethel. "No one knows what's going on, and no one at NMFS seems to care."

The explanation from NMFS was consistent and emphatic: The cod, perhaps in a move to maximize their chances of repopulating, had contracted into their core spawning areas in Ipswich, Mass, and Cape Cod Bays during March and April and, since they weren't harassed by nets, dug in for a spell, and thus were collected there when the area re-opened and the boats lowered their nets. There weren't any more cod than before; they were just all in one place, and the fishermen happened to be right on top of them.

2

On the summer solstice, a splendid day on which the still young leaves of New England shimmered gently in the high sun, I drove to Woods Hole through the clumpy, quiet green roll of Vermont and New Hampshire, savage Boston traffic, and finally out over the Bourne Bridge to salt air and water light. I found Jay Burnett in his office in the Age and Growth cottage. He was in running shorts and shoes and the same navy blue Henley shirt he often wore on the *Albatross,* talking on the phone, his feet propped up on his desk. His office, no more than an alcove, was perhaps all of 8 by 8 feet. He motioned for me to have a seat in a chair just outside the doorway (with no door) that separated his alcove from the long skinny room it had once been part of. The shelves of both rooms were stacked with pale brown cardboard boxes holding otoliths and scale samples from cruises past. They were labeled with heavy marker: bluefish, witch flounder, redfish.

He hung up, and after some small talk I asked him what he'd found on the spring Gulf cruise.

"Well, funny thing," he said. "We had perfect weather but saw scarcely *any* cod. We don't have all the numbers back yet, but it still looks like record-low numbers. I'm not sure how to reconcile that with what the fishermen are finding. Our official take—and this seemed consistent with what fish we found—is that what cod are in the Gulf are contracted into Ipswich, Mass, and Cape Cod Bays. And that certainly jibes with what we saw. That's about the only place we saw many fish. And we didn't hit any huge loads. But it was still early.

"It was interesting. Sam Novello went along—he's a dragger fisherman out of Gloucester, one of the guys I met at those meetings last summer— and he was very helpful identifying rock or ground cod versus the school cod. We didn't have a single tow where they were mixed. The red ones he called rock cod, or Gulf of Maine cod, and the silvery ones he said were the school fish, the ones that some of the guys are saying come from around the Cape. He said on a longer commercial tow they're almost always mixed. But on those short survey tows we always got just one or the other. So they really do seem to run separately. And the school cod were always right along the line described by the fishermen, running along the 50-fathom contour from the east flank of Stellwagen up to Jeffrey's.

"I sent otoliths from those fish along with a bunch of archived otoliths off to the guy at Sandy Hook. It'll be several months before we have definitive results. But for me, seeing those fish was very revealing. It really made me rethink this stock issue. I'm more convinced all the time that there's a stock on eastern Georges, one on western Georges, at least part of which slides up north along the 50-fathom contour into the Gulf, and another one that resides year-round in the western Gulf. Which is more or less the way the fishermen see it."

To his disappointment, he had learned that the otolith analysis at the Sandy Hook lab would take much longer than originally expected, probably until fall. He'd also had to pass off to the Island Institute the study fleet project, which he had hoped to use to start some cooperative research between NMFS and some of the fishermen he had met.

"They just didn't trust us enough," he explained. "We had this March meeting with the fishermen in Portland, and we hoped that some of them would agree to use some of the aid money from Clinton—this five million that's supposed to be both fisherman relief and research money—to pay them to collect some data for us. But the fishermen weren't game. They just don't trust NMFS enough at this point to cooperate with us on something like that. So we sort of turned that over to the Island Institute to see if they could get somewhere with it. At this point, I realized, we really need some mediating party between us and the fishermen. It's too bad. It would have been a good chance to get some cooperation going and get some good data. With a fleet of fishing boats bringing back good information, you could do all sorts of good work—tagging studies, detailed surveys of any sort of biological or fisheries question you can come up with. You could see what a cross-section of fisheries boats runs into and what sorts of operations they conduct and what effects those have. And you could get a much better handle on two critical statistics: a truer take on the catch-per-unit-effort of different boat sizes and gear types and, perhaps more important, the nature and amount of discards that fishing boats generate.

"That last issue is a real wild card. We really have no idea how many fish boats throw away. Some boats report huge discards and some none. We have to guess. But it's very important, especially with the low catch limits, to know how many fish are getting tossed overboard dead. Otherwise, we don't even know what the real impact of the regulations are."

I told him Goethel had described many boats discarding many cod.

"I know. I read his letter to the council. I can understand his disgust. And we've no idea whether that was an anomaly or if lots of fishermen are throwing tons of fish away. We don't know what the real mortality is."

Another setback—he seemed besieged by them—was a letter that Sam Novello, the fisherman who had been so helpful on the spring cruise, had written a couple of weeks earlier to the council and the local papers. Novello described how hard the crew on the survey cruise worked and how dedicated they seemed to be about doing good assessments (a big change in how most fishermen viewed NMFS staff), and he even said that the cruise had substantially changed his view of NMFS. But he also expressed dismay over what he saw as an assessment technique that was bound to miss fish: the big boat, the outmoded net, the short hauls.

There was, Jay recalled, a distinct moment on the spring cruise when Novello seemed to take on this skepticism. The *Albatross* happened to set near one of Novello's fishing friends in Mass Bay, close enough for Novello to recognize the boat, and Novello took the opportunity to go to the bridge and radio his friends and talk. Though they were towing parallel lines less than a mile apart, the other boat caught, in a tow just three times as long, about eight times as many cod as the *Albatross* did.

"Now, Sam knows that a half-mile in location can make a huge difference in what bottom you're fishing and whether you're going to catch fish. But he still couldn't get over that discrepancy in our catches. It seemed to puncture any faith he was gaining in what we were doing, that we could miss so many fish.

"So he went and wrote that letter. Some newspaper reporter jumped on it and used it to write a story that basically said, 'Fisherman says NMFS doesn't know what they're doing.' He quoted the letter very selectively. Used all the most critical stuff, called around and got some quotes to echo the criticisms, and ignored everything Sam wrote about how impressed he was with how hard everyone worked."

Despite all that, Burnett still believed it was best to take as many fishermen out on the cruises as possible. "There's just so much distrust. We're not going to change anything until that goes away and until they understand better what we're doing. We seem to insist on putting things in ways they can't understand. Not only can't our models *use* any information from fishermen, we can't even explain the models to them. No wonder they don't trust them! And now we've got these new so-called control rules we're supposed to use to meet some of the Sustainable Fisheries Act requirements, and they're even more arcane and complex. I almost hope the fishermen never see them.

"We do so little to make this science accessible. Jen Bubar, a fisherman out of Stonington, took me to task on this at one of the meetings last summer, saying that at so many of the council meetings or the scientific committee meetings you simply could not understand the scientists unless you were a scientist, too. She said fishermen shouldn't have to go to school to understand fisheries science. In a sense, she's right. We *should* work with these people. Bring them in for seminars if we have to, whatever it takes to make them really understand it."

He stopped speaking, and for a second I thought he had finished, either talked out or stricken with the thought that he had said too much.

"I've been disappointed in some of the people here. We do some great work here. I truly believe we put out some of the best fishery assessments in the world. But some of the inertia here is astounding. Like this stock issue. It's embarrassing we don't have a better grip on the stock distribution of this major species. We're working on a model fifty years old and we've got all this evidence that it's oversimplified. It's time to get that resolved. But I've run into a lot of trouble about this. There are people here who do not want to look at this stock distribution question.

"But if the otoliths shake out like I think they will, I think we'll be forced to. I think we'll have to do a full-blown distribution analysis. I wanted to do that myself. I could do it without too much trouble. But they said, No—no way was someone else going to do that."

He stopped a moment, shook his head, and laughed.

"So I'm not sure what I've done this last year or two other than get yelled at. Sometimes I don't think we've gotten anywhere. And there's this us-versus-them mentality that's setting in. A month or so ago, Mike Sissenwine debated Tom Brancaleone [a leader in the Gloucester fishing community] on Channel 4, and the whole thing was so frustrating and discouraging. There were all these people running around doing 'debate strategy,' trying to figure out how we could undermine Brancaleone. It was awful. We should be about the truth here. Not this us-against-them business. Sometimes I have to remind some of my colleagues that we're the National Marine *Fisheries* Service, not the Fish Service. We're supposed to manage the fishery, not just protect the fish. And to manage the fishery, you've got to work with the fishermen. We haven't been doing that.

"And you can have both. We've seen it. These fish populations can be fished awfully hard and still feed a lot of people when they're healthy.

"But what a mess it's become. It's a God-awful time."

3

The next day, wanting to see for myself how bad the discards were, I drove up the coast and checked into a hotel near the fishing pier on Hampton Beach. Remembering my exhausted, empty-stomach seasickness on my previous outing with Goethel, I walked down the boardwalk past the arcades and bars and drunken revelers and found a food store, where I bought two sandwiches, apples, a half-dozen bagels, and four bottles of water. Then I returned to my dank room, double-checked my supply of seasickness pills, and climbed into bed early. I opened *Fishes of the Gulf of Maine* and read once more the section on cod. One of the longer entries, it is also somewhat disjointed, with Bigelow's observations about the fishes' movement and stock distinctions, for instance, surfacing and then ending and then unexpectedly popping up again later—a pattern of appearance much like that of the fish themselves. I'd found before that this strange organization, along with the depth and variety of information the entry held, rewarded repeated readings; and I hoped that somewhere in those well-marked pages I would find the answer to the riddle of why Jay Burnett found so few fish and Dave Goethel so many. No go, however; I fell asleep as puzzled as ever.

At four-thirty the next morning, I walked to the pier and had my thermos filled with coffee by the gruff diner proprietor who fed the fishermen but hated the much more numerous tourists. (She had sold Goethel the tee-shirt he wore that day, the back of which read, "If it's called tourist season, why can't we shoot them?") I stood and looked at the predawn water while I ate a couple of bagels and sipped coffee. Goethel drove into the lot just before five, as promised. A few minutes later, we were in the *Ellen Diane* pulling out of the harbor and out the short canal toward Ipswich Bay.

Where the canal met the bay, we could see three anglers casting into the rip for stripers. I was glad to see it was a calm morning; the outgoing tide pulled little wakes around the navigation buoys, but otherwise the sea held only the gentlest of undulations. We passed hundreds of buoys marking lobster traps. This simplest and most accessible of fisheries had become more crowded than ever the past few years as many fishermen turned from the increasingly regulated and overfished ground- and midwater fisheries to lobstering, where so many of them had started their fishing careers. There were more people lobstering than ever, and the busier boats were handling more pots than ever before. So far the lobsters had held up, but NMFS was warning of a spectacular plunge if the number of pots wasn't soon curtailed.

Goethel did not want to talk about lobstering, though. He wanted to talk about cod. I asked him how the previous day had gone.

"About the same as the day before," he said. "Pretty much the same every day. About five hundred pounds of flounder, about a thousand pounds of cod. I'll tell you, I don't know how they could have missed this much cod. Even a few weeks ago I was willing to buy the idea that the cod were concentrating in the prime areas. But they were this thick in Mass Bay, and they've been this thick since the day I started fishing here, and they've been this thick in almost every piece of bottom along the way that might possibly hold cod. So we're talking about fifty miles of solid cod. I simply can't reconcile that number of fish with an overall population of only eight thousand metric tons.

"It's a disaster. You cannot avoid these fish no matter what you do. They're everywhere. It's awful. I hate it. Listen to me: I hate finding all these cod. This is what we've come to, where you hate finding cod. But the management makes it a disaster. I can't abide these discards. I've got a problem with killing fish for nothing. If I kill a fish, I want someone to use it. Right now the only ones using them are the slime eels [same as hagfish]. Slime eels!" He laughed a pained laugh. "We'll probably produce a huge bloom of slime eels now. Twenty years it'll be our main species. Slime eels and dogfish."

He looked out the window as if trying to control his anger or as if he were embarrassed about expressing a cynicism he had so long fought to avoid. I was a bit embarrassed myself for digging at such a painful subject. We'd talked about it numerous times, but Goethel expressed a sharpness and an anger now, a bitterness, that I hadn't seen before.

"I don't know how the science missed so many fish. But they did. They missed something. I can't prove it, but there's no doubt in my mind. They've missed something here."

He was quiet for a moment, and then he said, "You know what upsets me the most? No one from NMFS has come out here to see this. No observers. No boats. We have been complaining for three or four weeks, a month now, about all the fish we're killing, and no one from NMFS has come out on the fishing boats to see what we're talking about. I think they don't want to see it. They don't want to see how wrong they are. They'd rather fool themselves this policy is working. Because if they don't see it, then they can ignore all our 'anecdotal' stories about discards, undercount and ignore all the dead fish we're throwing out, and then decide this thirty-pound catch limit reduced fishing mortality below eight hundred metric tons. But whatever they're counting in the landings, there are ten times as many being thrown over.

"This is a disaster. It's the most disgusting thing I've ever seen. And it's set things back between NMFS and the fishermen years. It'll take a decade for the bitterness from this summer to wear off. They should be out here to see what's going on. If you want to know what's going on, you have to be out here. It's as simple as that."

The boat glided over the long swells. The rising sun cast a golden photographer's light over the satin water. Goethel peered ahead, then slowed the boat and prepared to make his first set. Everything paid out smoothly on this calm day, the net rolling into the water, and we were soon back in the cabin as we cruised south, the door open to let in the warming morning air. To starboard, the shore was easily visible, a low, undulating run of deep summer green and luminous yellow sand lit by the low sun. Here and there a stretch of shore buildings created some boxy clutter, and directly off our beam was a large structure that, with its towers flanked by rectangles, looked from here almost classical in its stature and symmetry: the Seabrook nuclear power plant.

For three hours we towed, talking, watching the water, monitoring the radio. Most of the radio chatter Dave ignored, but he talked now and then with his friend Carl, another small-boat dragger who often fished in tandem with Dave, working the same area and comparing notes. The water remained a glistening sheet.

I asked him if he'd break even this year.

"Well," he said, opening a container of fruit salad. "Year's still early. But right now I have seventeen thousand less in the bank than I did on January first. I had to replace the prop and the prop shaft in May. That was about five grand. The winches, those were about three. I put in new doors that open the net wider for flounder, since I'm working so much on flounder now and not cod. Those were another thirty-five hundred. And I still need to get new net cables, as my old ones are getting old, and to replace the sweeper on my groundfish net. I can put some of that stuff off, but it catches up to you. Something breaks and you can't even fish. So I don't know. It doesn't look too good at this point. My wife keeps asking me why I put money into this. I tell her I have no choice. We'll hang in there as long as we can."

At nine, he pulled on his bib overalls, rubber boots and gloves, and a tennis hat and went out to retrieve the net. Soon the buoys and the top net cord were bounding along in the water behind, and the usual seagulls materialized to pick at the net. As the belly of the net reached the reel, Goethel stopped it every few feet to shake out the fish, mostly flounder, that were tangled in the mesh; they tumbled down the loose sleeve of the net toward the section still in the water. Soon the tighter, heavier mesh of the cod end came over the stern. The bag looked modestly but pleasingly round and full, a ball about the size of a washing machine. He worked the reel until the load was suspended a few feet over the fish deck, then with a rubber mallet knocked loose the pin that held the puckerstring in place. The fish tumbled out. Most were flounder, sliding in layers of brownish gray. Among them were a few sculpins, three or four grim-looking dogfish, some cusk, a dozen or so lobsters, and perhaps fifty cod. Goethel threw the saltwater hose onto the catch and we bent over the pile and worked to get the cod overboard as quickly as possible: a grab of the fish by the tail or the lip, a hand under the belly, a lift over the gunwale and into the water. This catch-and-release fishing felt familiar to me from my hundreds of days on trout streams, and for once on a fishing boat my inclination to carefully tend each fish seemed not so effete and out of place. Even so, I quickly realized I couldn't handle them too carefully; speed was of the essence. So I took to working faster, as Goethel was, grabbing a fish, handling it with reasonable care but getting it overboard quickly. Now and then, one would immediately dart off. But most of them, in the brief look I gave them as they hit the water, either floated or rolled over, lifeless.

Finally, we had them all over but for three Goethel set in a box to make his limit of 30 pounds. I stood up and looked about. Behind us, floating on the surface, were a good two dozen fish, white bellies up.

"How many of the others you think make it?" I asked. "Half?"

"Maybe that," he said.

For a moment, neither of us had anything to say.

"How much you figure we threw over?" I asked.

"That was about four hundred pounds," he said, his voice tight. "Probably two or three hundred pounds dead."

He reached down to his feet and grabbed two fish I had not yet noticed and held them up by their tails.

"I wanted to show you these," he said.

They had once been magnificent fish, almost 3 feet long and thigh-thick. Now they were carcasses: gray sacs of old flesh, the eye sockets sunken and the eyes eaten out, some of the bellies chewed away, ribs showing, fins worn away, everything slack, the fish consumed inside out by hagfish and sand fleas. Goethel and other fishermen had told me about such fish. They were catching more and more of them as summer went on. They were discards from previous trips, tossed overboard as we had just done, and they now so littered the floor of the Gulf that they were being scooped up a second time.

"This is where we've come," he said, looking at the fish. "This is our legacy."

With a gentleness he had lacked time for with the live ones, he lifted the fish over the gunwale and slipped them into the water. They sank into the clear sea, grew dim, and disappeared.

EPILOGUE

Late in the autumn of 1999, Jay Burnett finally received the results on the cod otoliths he had sent to the Sandy Hook lab for stock distribution analysis. He had hoped the tests would shed light on the question of how many cod stocks moved through the waters of New England and in particular on whether NMFS's two-stock picture might need revision. He was disappointed, to put it mildly, that the chemical analysis of the otoliths showed, as he put it, "absolutely nothing. They were completely inconclusive." Something, the chemist explained, had gone amiss: Either the sample batches were too small, or they had been improperly classified, or something was wrong with the equipment, or the notoriously finicky nature of the procedure and method had made itself manifest. Whatever the cause, the results were too fuzzy to support any particular conclusions. In neither batch of otoliths Burnett sent down—not the relatively fresh batch he had collected with Sam Novello from Georges and the Gulf on the spring 1999 *Albatross* cruises nor the archived otoliths from cruises past, which came from all over Georges and the Gulf—did the testing show substantial differences in chemical signatures between stocks. The chemist even ran a second batch, hoping it would be different, but got the same results. As far as the testing showed, all the otoliths came from one big stock, with no distinction even between Georges Bank and Gulf of Maine fish. Everyone agreed it had to be a testing problem. The only way to correct it, if everyone was up for the work and expense, was to assemble two entirely new batches drawn from all over Georges and the Gulf and do everything all over again.

"So we don't know what it means," said Jay after he got the news. "You can say it means anything you want to because it doesn't mean anything. The results probably support any theory—or no theory—you want to put forth. I suspect those defending the two-stock theory are happy, because it doesn't do anything to threaten that. Everybody acknowledges we need more information. But we have no budget. The best we can do to look into these things is list them among our desired research projects and hope some Ph.D. candidate picks it up."

Though he was disappointed that his own in-house efforts had hit the rocks, Burnett could find hope that similar projects might find support through an additional $6 million in federal aid that Congress had recently allocated specifically for cooperative research projects in New England involving scientists and fishermen. The priorities and guidelines for that research would be set by a new advisory steering committee composed of fishermen and scientists appointed by the New England Fishery Management Council. The money and the new steering committee were in essence an official continuation, at least in spirit, of the work that Burnett, the Island Institute, the Gulf of Maine Aquarium, and others had done over the previous year or two to stimulate more dialogue and research involving fishermen and scientists, and which was now being carried on in various forms by those organizations and a growing list of others, such as the Gulf of Maine Fishermen's Alliance, the University of Maine's Darling Marine Center, and the Northeast Consortium, a new research cooperative made up of researchers from the University of Maine, the University of New Hampshire, the Massachusetts Institute of Technology, and the Woods Hole Oceanographic Institution. Things were finally stirring. The growing discord over assessments, and in particular the cod-discard disaster the previous summer, had got everyone's attention. If the distrust between fishermen and the Science Center and the negative inertia within the Science Center still weighed too heavily to allow NMFS to directly carry out such research—well, figured Burnett, maybe that was just how it needed to be for now.

"Maybe," he said, with what seemed a mixture of genuine hope and the vague awareness that he might be fooling himself, "we just need a layer between us and the fishermen for now. Maybe this is just how things need to get started."

As it happened, the scientific steering committee that formed in September of 1999 included Dave Goethel. Goethel had narrowly missed being appointed to the New England Fishery Management Council earlier that summer; he was reportedly near or at the top of the list New Hampshire governor Jeanne Shaheen submitted to the Commerce Department but was passed over. At least one person I talked to was glad that Goethel was put on the research steering committee instead, where he could bring his experience and knowledge more directly to bear.

Goethel joined the committee still angry about the summer's discard fiasco but glad to have a chance to work on what he saw as the root cause.[*] He became increasingly certain that NMFS was getting something fundamentally wrong in its Gulf cod assessments. He was determined to push for studies that would shed more light on issues of stock identification and the numbers, movement, and concentration of cod in the Gulf of Maine.

The rest of the committee was made up of present and past NMFS assessment scientists, scientists from academe and private institutions, four fishery council members (all fishermen), and fisherman and former fishery council member Frank Mirarchi, who worked the Gulf with a 60-foot dragger out of Scituate, Massachusetts, and shared many of Goethel's experiences and ideas. The group was to designate a list of research priorities, call for and review research proposals, then forward a recommended list of projects to the council and the Science Center, which would accept or amend the recommended list.

With roughly equal representation from the industry and scientific worlds and an order from the council to decide things by consensus, the committee entered its long winter of deliberations knowing it would have to get along to accomplish anything. And after the first, rather uneasy

---

[*]In July 1999, responding to the outcry among fishermen and a request from the New England Fishery Management Council, NMFS had raised the daily cod-take limit in the Gulf of Maine to 100 pounds and then to 200 pounds. While this reduced discard somewhat and allowed fishermen to make a bit more money, most Gulf fishermen ended the summer firmly convinced that NMFS's assessments of Gulf cod were unreliable.

meeting or two, everyone seemed pleased at the level of knowledge and rapport.

"We seem to have some of the better people from both camps," said Goethel. "At first there was a lot of nervousness and some condescension. But once the scientists saw we could keep up, we started to get some good work done. Granted, everybody's got one or two ideas they don't want to give up on, and that may hold us up. But for now we're at least exchanging information, even if it pains us at times. That's an improvement. And if we get some useful projects that answer some of the questions that have been ignored for so long, that'll be real progress." By early 2000, Goethel grew guardedly optimistic that the committee would call for, among other measures, tagging studies and some form of study fleet that could help provide detailed stock definitions, better counts of actual takes and discard rates, and perhaps a beginning on collecting more fine-grained census data from industry boats to complement NMFS's twice-yearly surveys.

Whether and how quickly the committee's priorities and recommendations would garner NMFS approval and become research projects—and whether the projects would fill the gaps in science and dialogue that undermined the management of the New England fishery—depended on many things the committee could not control. Not least among these were continued funding and a sustained commitment from everyone involved. More fishermen would have to see the value of working with an agency they didn't trust, and the most resistant scientists at the Northeast Fisheries Science Center would have to change some of their methods and thinking. Even if good cooperative projects started generating data, some way would have to be found to mesh those findings with those of NMFS if the new information wasn't to become ammo in a redoubled spat over numbers.

Progress would require a major shift in thinking on both sides. As Frank Mirarchi put it, "We're just beginning to deal with something we should have been tending to for over a decade. It's ten years now we've been yelling at each other, and while we've been yelling the fish have been going down the drain. Now a few people have stopped yelling and started talking. But a lot more has to happen before we begin to do things better. We've got an awfully big boat to turn around."

Goethel, meanwhile, continued to nimbly work his own rather small boat. In January 2000, the New England Fishery Management Council passed a lightly amended version of the previous year's spring and early-summer cod closure in the Gulf, practically guaranteeing that Goethel would get little income from groundfish. To close the gap, he fished especially hard that winter during shrimp season, working almost every day open to him, even though that meant going out on days rougher than he ordinarily liked to work. He found a crew member, one of the few people not already employed in New England's thriving shore economy, and every morning woke at 3:30, drove to the crewman's apartment, banged the door to get him out of bed, tried to ignore the beer cans visible when the guy blearily answered the door, and then drove to the dock with him and set out to chase shrimp clouds.

He didn't have much company on the water. NMFS had predicted a poor shrimp season, so most of the other small-boat captains declined to rig for shrimp that winter. They figured it wasn't worth sailing in the rough, freezing weather of February and March just to break even. Goethel, however, saw it as either fish or start sliding a slippery slope that ended on shore.

"Between not shrimping this winter and staying home during the cod season, a lot of the guys are taking the whole first half of the year off," he said. "How are they going to make up for that? Some are working on shore. My guess is quite a few of those guys will hang it up. Their ties with the fishing community will erode. They stay ashore now, what's going to get them back out? There is no reason to think things will improve for at least a couple of years. The regulations aren't going to lighten up anytime soon. And working ashore, the reality of getting up every morning at seven instead of four will appeal to them more and more. It'll get harder and harder to go to the dock. It's like getting back on the horse that threw you. I've always thought if I stopped a year, I'd never go back. Way it is now, I get up at four every morning whether I'm fishing or not. It's automatic. If I got up at seven for a year, I don't know if I could start doing it again. It's just too hard."

At least for that winter, getting up at four paid off. Goethel rigged his shrimp net in January worrying that he would finally go broke that year. For the first time in a while, though, everything went his way: Despite NMFS's predicted shrimp scarcity, Goethel found shrimp in numbers he had not seen in more than a decade. "We fished some hard weather in that stretch," he said later, "weather that stunned everyone. But if you can make one tow and catch two thousand pounds of shrimp, you do it, even if it's five below and blowing." With the scarcity of boats making it a seller's market, he sold his tons of shrimp for close to a dollar a pound. By the time the season closed, in late March, he had stashed enough money to get him clear through the cod closures of April and May and June and into the summer, even if he didn't go out at all during that time. The shrimp saved him.

"Why all those shrimp were there when NMFS said they wouldn't be," he said, "I don't know. But on this one, I'm not complaining."

NOTES

Prologue

1. Mark Kurlansky, *Cod: A Biography of the Fish That Changed the World* (New York: Walker and Co., 1997).
2. David Dobbs and Richard Ober, *The Northern Forest* (White River Junction, VT: Chelsea Green, 1995).

Part I

1. Henry Bigelow and William Schroeder, *Fishes of the Gulf of Maine*, U.S. Fish and Wildlife Service Bulletin 74 (vol. 53) (Washington, DC: U.S. Government Printing Office, 1953).
2. Henry B. Bigelow, "A Developing View-Point in Oceanography," *Science* 71 (January 24, 1930): 84–89.
3. Henry Bigelow to Alexander Agassiz, March 10, 1902, in Harvard Museum of Comparative Zoology's Ernst Mayr Library, Special Collection bAg 99–101.
4. Henry B. Bigelow, *Memories of a Long and Active Life* (Cambridge, MA: Cosmos Press, 1964), p. 4.
5. Ibid., p. 7.
6. Henry B. Bigelow, "Birds of the Northeastern Coast of Labrador," *The Auk* 19 (1902): 24–31.
7. Sir John Murray, "Alexander Agassiz: His Life and Scientific Work," *Science* 33, no. 858 (June 9, 1911): 880.
8. George R. Agassiz, *Letters and Recollections of Alexander Agassiz* (Boston: Houghton Mifflin, 1913), pp. 395–396.
9. Henry Bigelow, "A Developing View-Point in Oceanography," *Science* 71 (January 24, 1930), p. 86.

10. Bigelow, *Memories,* p. 20.
11. The work on medusae is Henry B. Bigelow, "The Medusae," *Memoirs of the Museum of Comparative Zoology, Harvard College* 37 (1909). The quote is from Bigelow, *Memories,* p. 21.
12. Bigelow, *Memories,* p. 23.
13. Columbus Iselin, *Oceanus* 19, no. 2 (July 1968): 8.
14. Henry B. Bigelow, "Explorations in the Gulf of Maine, July and August, 1912, by the U.S. Fisheries Schooner *Grampus," Bulletin of the Museum of Comparative Zoology* 63, no. 2 (February 1914): 88.
15. Ibid., p. 234.
16. One good bathymetric chart is Elazar Uchupi's "Georges Bank and Vicinity" (Woods Hole, MA: Woods Hole Oceanographic Institution, 1987). Some of the oceanographic features of the map on the endpapers of this book are taken from Uchupi's chart.
17. Bigelow, "Explorations in the Gulf of Maine, July and August, 1912," p. 100.
18. Ibid., p. 34.
19. Ibid., p. 98.
20. Bigelow to H. F. Moore, April 13, 1920, Papers of Henry B. Bigelow (HUG4212.xx), Harvard University Archives.
21. Henry Bigelow, "Physical Oceanography of the Gulf of Maine," *Bulletin of the U.S. Bureau of Fisheries* 40, no. 2 (1927): 511–1027.
22. Henry Bigelow, "Exploration of the Waters of the Gulf of Maine," *Geographical Review* 18 (1928): 250.
23. A. G. Huntsman, "The Importance of Tidal and Other Oscillations in Ocean Circulation," *Transcripts of the Royal Society of Canada,* series 3, vol. 17 (1923), sec. 5, pp. 15–20.
24. As of early 2000, these animations could be downloaded from Professor Dan Lynch's web site at <http://www-nml.dartmouth.edu/Publications/internal_reports/NML-98-7>.
25. Bigelow and Schroeder, *Fishes of the Gulf of Maine,* p. 340.
26. Ibid., p. 341. The Wulff citation is from the *International Game Fish Association Yearbook,* 1945, p. 65.
27. Bigelow and Schroeder, *Fishes of the Gulf of Maine,* p. 11.
28. Ibid., p. 529.
29. Ibid., p. 211.
30. Ibid., p. 213.
31. Ibid., p. 254.

Part II

1. Henry Bigelow and William Schroeder, *Fishes of the Gulf of Maine*, U.S. Fish and Wildlife Service Bulletin 74 (vol. 53) (Washington, DC: U.S. Government Printing Office, 1953), p. 60.

2. National Research Council, *Review of Northeast Fishery Stock Assessments* (Washington, DC: National Academy Press, 1998). Also available online at <http://www.nas.org>.

3. National Research Council, *Improving Fish Stock Assessments* (Washington, DC: National Academy Press, 1998). Also available online at <http://www.nas.org>.

4. Bigelow and Schroeder, *Fishes of the Gulf of Maine*, p. 48.

5. Daniel Pauly, V. Christensen, J. Dalsgaard, R. Froese, and F. Torres Jr., "Fishing Down Marine Food Webs," *Science* 279 (1998): 860–863.

6. Personal communication, Jason Link, NMFS Northeast Fishery Center Food Habits Division, November 9, 1999. Data can be viewed at <http://www.nefsc.nmfs.gov/pbio/fwdp/projects.htm#4>.

7. Pauly ct al., "Fishing Down Marine Food Webs," p. 862.

8. Tim Smith, *Scaling Fisheries: The Science of Measuring the Effects of Fishing, 1855–1955* (New York: Cambridge University Press, 1994), p. 2.

9. Georg Ossian Sars, "On the Spawning and Development of the Cod-Fish," *Report of the United States Fish Commission* 3 (1876): 213–222, as cited in Smith, *Scaling Fisheries.*

Part III

1. Much of this information on the growth of the New England fleet is from Andrew Kitts, Eric Thunberg, and G.T.K. Sheppard, "The Northeast Groundfish Fishery Buyout Program," in Stephen H. Clarke, ed., *Status of the Fishery Resources off the Northeastern United States for 1998,* NOAA Technical Memorandum NMFS-NE-115. Available online at <http://www.nefsc.nmfs.gov/sos/tables/figure17.html>.

2. Mark Kurlansky, *Cod: A Biography of the Fish That Changed the World* (New York: Walker and Co., 1997).

3. Walter H. Rich, *Fishing Grounds of the Gulf of Maine* (1929; reprint, Augusta: Maine Department of Marine Resources, 1983).

4. Henry Bigelow and William Schroeder, *Fishes of the Gulf of Maine*, U.S. Fish and Wildlife Service Bulletin 74 (vol. 53) (Washington, DC: U.S. Government Printing Office, 1953), pp. 192–193.

Part IV

1. Henry Bigelow and William Schroeder, *Fishes of the Gulf of Maine,* U.S. Fish and Wildlife Service Bulletin 74 (vol. 53) (Washington, DC: U.S. Government Printing Office, 1953), p. 191.
2. Edward Ames, *Cod and Haddock Spawning Grounds of the Gulf of Maine* (Rockland, ME: Island Institute, 1997).
3. National Research Council, *Review of Northeast Fishery Stock Assessments* (Washington, DC: National Academy Press, 1998), p. 5.
4. D. E. Ruzzante, C. T. Taggart, and D. Cook, "A Review of Evidence for Genetic Structure of Cod *(Gadus morhua)* Populations in the NW Atlantic and Population Affinities of Larval Cod off Newfoundland and the Gulf of St. Lawrence," *Fisheries Research* 43 (1999): 79–97. Virtually all the other articles in this special issue on stock identification are also relevant.

Part V

1. Henry Bigelow and William Schroeder, *Fishes of the Gulf of Maine,* U.S. Fish and Wildlife Service Bulletin 74 (vol. 53) (Washington, DC: U.S. Government Printing Office, 1953), p. 90.

BIBLIOGRAPHY

Agassiz, Alexander. *Letters and Recollections of Alexander Agassiz, with a Sketch of His Life and Work.* Boston: Houghton Mifflin, 1913.

Ames, Edward. *Cod and Haddock Spawning Grounds of the Gulf of Maine.* Rockland, ME: Island Institute, 1997.

Backus, Richard, ed. *Georges Bank.* Cambridge, MA: MIT Press, 1987.

Begg, Gavin, ed. *Fisheries Research* 43 (1999) (special issue on stock identification).

Bigelow, Henry B. Papers of Henry Bryant Bigelow, 1906–1964. Harvard University Archives, Cambridge, Mass.

———. "Explorations in the Gulf of Maine, July and August, 1912, by the U.S. Fisheries Schooner *Grampus.*" *Bulletin of the Museum of Comparative Zoology* 63, no. 2 (February 1914): 29–145.

———. *Plankton of the Offshore Waters of the Gulf of Maine.* Washington, DC: U.S. Government Printing Office, 1926.

———. "Physical Oceanography of the Gulf of Maine." *Bulletin of the U.S. Bureau of Fisheries* 40, no. 2 (1927): 511–1027.

———. "Exploration of the Waters of the Gulf of Maine." *Geographical Review* 18 (1928): 232–260.

———. "A Developing View-Point in Oceanography." *Science* 71 (January 24, 1930): 84–89.

———. *Oceanography: Its Scope, Problems, and Economic Importance.* Boston: Houghton Mifflin, 1931.

———. *Memories of a Long and Active Life.* Cambridge, MA: Cosmos Press, 1964.

Bigelow, Henry B., and William C. Schroeder. *Fishes of the Gulf of Maine.* U.S. Fish and Wildlife Service Bulletin 74 (vol. 53). Washington, DC: U.S. Government Printing Office, 1953.

Boreman, B. S. Nakashima, and R. L. Kendall, eds. *Northwest Atlantic Groundfish: Perspectives on a Fishery Collapse*. Bethesda, MD: American Fisheries Society, 1997.

Brosco, Jeffrey P. "Henry Bryant Bigelow, the U.S. Bureau of Fisheries, and Intensive Area Study." *Social Studies of Science* 19 (1989): 239–264.

Carey, Richard A. *Against the Tide: The Fate of the New England Fisherman*. Boston: Houghton Mifflin, 1999.

Clarke, Stephen H., ed. *Status of the Fishery Resources off the Northeastern United States, 1998*. NOAA Technical Memorandum NMFS-NE-115. Woods Hole, MA: Northeast Fisheries Science Center, 1998. Also available online at <http://www.wh.whoi.edu/sos/>.

Conkling, Philip, ed. *From Cape Cod to the Bay of Fundy: An Environmental Atlas of the Gulf of Maine*. Cambridge, MA: MIT Press, 1995.

Crestin, David. "Federal Regulation of Fisheries." Paper presented at the Connecticut College Center for Conservation Biology and Environmental Studies Conference on the History, Status, and Future of the New England Offshore Fishery, Connecticut College, New London, Conn., April 16–17, 1999.

Doeringer, Peter B. *The New England Fishing Economy: Jobs, Income, and Kinship*. Amherst: University of Massachusetts Press, 1986.

Fordham, S. *New England Groundfish: From Glory to Grief. A Portrait of America's Most Devastated Fishery*. Washington, DC: Center for Marine Conservation, 1996.

Hahn, Jan, ed. *Oceanus* 14, no. 2 (July 1968) (special issue on Henry Bigelow).

Herdman, William A. *Founders of Oceanography: An Introduction to the Science of the Sea*. London: Edward Arnold, 1923.

Kurlansky, Mark. *Cod: A Biography of the Fish That Changed the World*. New York: Walker and Co., 1997.

Lough, Greg, W. G. Smith, F. E. Werner, J. W. Loder, F. H. Page, C. G. Hannah, C. E. Naimie, R. I. Perry, M. Sinclair, and D. R. Lynch. "Influence of Wind-Driven Advection on Interannual Variability in Cod Egg and Larval Distributions on Georges Bank: 1982 vs. 1985." *ICES Marine Science Symposium* 198 (1994): 356–378.

Lough, R. Gregory, and D. C. Potter. "Vertical Distribution Patterns and Diel Migrations of Larval and Juvenile Haddock *Melanogrammus aeglefinus* and Atlantic Cod *Gadus morhua* on Georges Bank." *Fishery Bulletin* 91, no. 2 (1993): 281–303.

Lynch, Daniel, J. T. C. Ip, C. E. Naimie, and F. E. Werner. "Comprehensive Coastal Circulation Model with Application to the Gulf of Maine." *Continental Shelf Research* 16, no. 7 (1996): 875–906.

Lynch, Daniel R., W. G. Gentleman, D. McGillicutty, and Cabell S. Davis. "Biophysical Simulations of *Calanus finmarchius* Population Dynamics in the Gulf of Maine." Draft of a paper prepared for the Marine Ecology Progress Series, 1998.

Murray, John. "Alexander Agassiz: His Life and Scientific Work." *Science* 33, no. 858 (June 9, 1911): 873–887.

Myers, R. A., N. J. Barrowman, J. A. Hutchings, and A. A. Rosenberg. "Population Dynamics of Exploited Fish Stocks at Low Population Levels." *Science* 269 (1995): 1106–1108.

National Research Council. *An Assessment of the Atlantic Bluefin Tuna.* Washington, DC: National Academy Press, 1994.

———. *Improving Fish Stock Assessments.* Washington, DC: National Academy Press, 1998.

———. *Review of Northeast Fishery Stock Assessments.* Washington, DC: National Academy Press, 1998.

Oceanography: The Past. Proceedings of the Third International Congress on the History of Oceanography. New York: Springer-Verlag, 1980.

Pauly, Daniel, V. Christensen, J. Dalsgaard, R. Froese, and F. Torres Jr. "Fishing Down Marine Food Webs." *Science* 279 (February 6, 1998): 860–863.

Redfield, Alfred C. "Henry Bryant Bigelow." *Biographical Memoirs* 48 (51–80).

Rich, Walter H. *Fishing Grounds of the Gulf of Maine.* Reprint, Augusta: Maine Department of Marine Resources, 1983.

Safina, Carl. *Song for the Blue Ocean.* New York: Holt, 1998.

Schlee, Susan. "The R/V *Atlantis* and Her First Oceanographic Institution." In C. P. Idyll, ed., *Exploring the Ocean World: A History of Oceanography.* New York: Crowell, 1972.

———. *The Edge of an Unfamiliar World: A History of Oceanography.* New York: E. P. Dutton, 1973.

Smith, Tim D. *Scaling Fisheries: The Science of Measuring the Effects of Fishing, 1855–1955.* New York: Cambridge University Press, 1994.

Thoreau, Henry. *Cape Cod.* Princeton, NJ: Princeton University Press, 1993.

Warner, William. *Distant Water: The Fate of the North Atlantic Fisherman.* Boston: Little, Brown, 1977.

Werner, Francisco, F. H. Page, D. R. Lynch, J. W. Loder, R. G. Lough, R. I. Perry, D. A. Greenberg, and M. M. Sinclair. "Influences of Mean Advection and Simple Behavior on the Distribution of Cod and Haddock Early Life Stages on Georges Bank." *Fisheries Oceanography* 2, no. 2 (1993): 43–64.

Whynott, Douglas. *Giant Bluefin.* New York: Farrar, Straus and Giroux, 1995.

Wilson, Douglas C. "Fisheries Science Collaborations: The Critical Role of the Community." Paper presented at the Conference on Holistic Management and the Role of Fisheries and Mariculture in the Coastal Community, Tjärnö Marine Biological Laboratory, Sweden, November 11–12, 1999.

INDEX